Quantum Entanglement Engineering and Applications

IOP Series in Coherent Sources, Quantum Fundamentals, and Applications

About the Editor

F J Duarte is a laser physicist based in Western New York, USA. His career has expanded three continents while contributing in the academic, industrial and defense sectors. Duarte is editor/author of 15 laser optics books and sole author of three books: *Tunable Laser Optics, Quantum Optics for Engineers*, and *Fundamentals of Quantum Entanglement.* Duarte has made original contributions in the fields of coherent imaging, directed energy, high-power tunable lasers, laser metrology, liquid and solid-state organic gain media, narrow-linewidth tunable laser oscillators, organic semiconductor coherent emission, *N*-slit quantum interferometry, polarization rotation, quantum entanglement, and space-to-space secure interferometric communications. He is also the author of the generalized multiple-prism grating dispersion theory and pioneered the use of Dirac's quantum notation in *N*-slit interferometry and classical optics. His contributions have found applications in numerous fields, including astronomical instrumentation, dispersive optics, femtosecond laser microscopy, geodesics, gravitational lensing, heat transfer, laser isotope separation, laser medicine, laser pulse compression, laser spectroscopy, mathematical transforms, nonlinear optics, polarization optics, and tunable diode laser design. Duarte was elected Fellow of the Australian Institute of Physics in 1987 and Fellow of the Optical Society of America in 1993. He has received various recognitions, including the *Paul F Foreman Engineering Excellence Award* and the *David Richardson Medal* from the Optical Society.

Coherent Sources, Quantum Fundamentals, and Applications

Since its discovery, the laser, has found innumerable applications from astronomy to… zoology. Subsequently, we have also become familiar with additional sources of coherent radiation such as the free electron laser, optical parametric oscillators, and coherent interferometric emitters. The aim of this book Series in Coherent Sources, Quantum Fundamentals, and Applications is to explore and explain the physics and technology of widely applied sources of coherent radiation and to match them with utilitarian and cutting-edge scientific applications. Coherent sources of interest are those that offer advantages in particular emission characteristics areas such as broad tunability, high spectral coherence, high energy, or high power. An additional area of inclusion are those coherent sources capable of high performance in the miniaturized realm. Understanding of quantum fundamentals can lead to new and better coherent sources and unimagined scientific and technological applications. Application areas of interest include the industrial, commercial, and medical sectors. Also, particular attention is given to scientific applications with a bright future such as coherent spectroscopy, astronomy, biophotonics, space communications, space interferometry, quantum entanglement, and quantum interference.

Publishing benefits
Authors are encouraged to take advantage of the features made possible by electronic publication to enhance the reader experience through the use of colour, animation and video, and incorporating supplementary files in their work.

Do you have an idea of a book that you'd like to explore?
For further information and details of submitting book proposals, see iopscience.org/books or contact Ashley Gasque at ashley.gasque@iop.org.

Quantum Entanglement Engineering and Applications

F J Duarte and T S Taylor

IOP Publishing, Bristol, UK

ISBN 978-0-7503-3407-5 (ebook)
ISBN 978-0-7503-3405-1 (print)
ISBN 978-0-7503-3408-2 (myPrint)
ISBN 978-0-7503-3406-8 (mobi)

DOI 10.1088/978-0-7503-3407-5

Version: 20210701

IOP ebooks

British Library Cataloguing-in-Publication Data: A catalogue record for this book is available from the British Library.

Published by IOP Publishing, wholly owned by The Institute of Physics, London

IOP Publishing, Temple Circus, Temple Way, Bristol, BS1 6HG, UK

US Office: IOP Publishing, Inc., 190 North Independence Mall West, Suite 601, Philadelphia, PA 19106, USA

In memory of John Clive Ward (1924–2000)

Contents

Series preface

The laser, a bright child of the quantum era and wondrous emitter of coherent radiation, has offered humankind solutions in a vast spectrum of disciplines. Fields from astronomy to medicine have benefited enormously from the laser. Concurrently, additional sources of coherent radiation such as the free electron laser, optical parametric oscillators, and interferometric emitters, have increased the freedom of choice for scientists and engineers searching for the perfect match between source and application. The aim of this book Series in *Coherent Sources, Quantum Fundamentals, and Applications* is to explore and explain the physics and technology of widely applied sources of coherent radiation and to match them with utilitarian and cutting-edge scientific applications. Coherent sources of interest are those that offer particular advantages in specific emission characteristics areas such as broad tunability, high energy, and high power. An additional area of inclusion is those coherent sources capable of high performance in the miniaturized realm. Understanding of quantum fundamentals can lead to new and better coherent sources and unimagined scientific and technological applications. Application areas of interest include the industrial, commercial, and medical sectors. Also, particular attention is given to scientific applications with a bright future such as coherent spectroscopy, astronomy, biophotonics, space communications, space interferometry, quantum entanglement, and quantum interference.

Book preface

Quantum Entanglement Engineering and Applications is a book-textbook comprised of 10 chapters and one appendix. All chapters include a set of problems. A principal motivating factor behind the creation of this book is to provide experimental physicists, scientists, and engineers, outside the specialty of quantum optics, with a transparent, direct, and practical guide to the field of quantum entanglement. In this regard, *Quantum Entanglement Engineering and Applications* provides a solid background on Dirac's *bra ket* notation that is essential to the cohesive and unified description of quantum interferometry and quantum entanglement via the Pryce–Ward probability amplitude $|\psi\rangle = (|x_1, y_2\rangle \pm |y_1, x_2\rangle)$, also known as Bell's states, that applies to two ($n = 2$) quanta propagating via two ($N = 2$) channels. This interferometric framework is then extended to generalized probability amplitudes applicable to $n = N = 2^1, 2^2, 2^3, \ldots, 2^r$ and $n = N = 3, 6, 9$ situations. It is all the same physics! The book then emphasizes the importance of *indistinguishability* prior to describing quantum interferometry via Dirac's notation. This description considers a wide range of interferometers including the Hanbury Brown–Twist, HOM, Mach–Zehnder, Michelson, N-slit, and Sagnac interferometers. The focus attention then turns to vector and matrix techniques necessary to describe quantum entanglement in matrix form. In this regard the nexus between $|\psi\rangle = (|x_1, y_2\rangle \pm |y_1, x_2\rangle)$ with Pauli matrices and the Hadamard matrix is described in an explicit and transparent approach. In a unique chapter, five physics avenues to arrive to the Pryce–Ward probability amplitude are provided. In its second phase *Quantum Entanglement Engineering and Applications* turns its attention to applications including quantum cryptography, quantum computing, quantum teleportation, and quantum metrology. It does so with little use of words and ample use of transparent mathematics. A separate chapter is dedicated to space-to-space quantum communications. The book concludes with an in-depth discussion on the pragmatic approach to quantum mechanics and quantum entanglement based on the documented views and perspectives of Dirac, Dyson, Feynman, Lamb, and Ward. This chapter includes a frank discussion on EPR's criticisms, Bell's theorem, von Neumann's opinion, and hidden variable theories. Reinforcing Feynman's vision, this pragmatic perspective on quantum entanglement, and quantum mechanics, is free of paradoxes.

Acknowledgements

One of us (FJD) would like to acknowledge support from *Interferometric Optics* during the gestation of this work.

Author biographies

F J Duarte

F J Duarte is a Chilean-born laser physicist based in Western New York, USA. He graduated with *First Class Honours* in physics from Macquarie University where he was a student of the quantum and particle physicist J C Ward. At Macquarie he also completed his PhD research on optically-pumped molecular lasers. He then became a Postdoctoral Fellow at the University of New South Wales, where he built UV narrow-linewidth tunable lasers for high-resolution IR–UV double-resonance spectroscopy. Duarte has worked and contributed professionally in the academic, industrial, and defense sectors and has practiced physics in Australia, the Americas, and Europe. He is the author of numerous archival physics papers and an inventor in the fields of lasers and optics. Duarte is editor/author of 15 scholarly books including *Dye Laser Principles*, *High-Power Dye Lasers*, *Organic Lasers and Organic Photonics*, *Tunable Laser Applications*, and *Tunable Lasers Handbook*. He is sole author of *Tunable Laser Optics*, *Quantum Optics for Engineers*, and *Fundamentals of Quantum Entanglement*. Duarte has made original contributions in the fields of coherent imaging, directed energy, high-power tunable lasers, laser metrology, liquid and solid-state organic gain media, narrow-linewidth tunable laser oscillators, organic semiconductor coherent emission, *N*-slit quantum interferometry, polarization rotation, quantum entanglement, and space-to-space secure interferometric communications. He is also the author of the generalized multiple-prism grating dispersion theory and pioneered the use of Dirac's quantum notation in *N*-slit interferometry and classical optics. His contributions have found applications in astronomical instrumentation, dispersive optics, femtosecond laser microscopy, geodesics, gravitational lensing, heat transfer, laser isotope separation, laser medicine, laser pulse compression, laser spectroscopy, mathematical transforms, nonlinear optics, polarization optics, and tunable diode laser design. Current interests include semiconductor organic lasers, tunable laser physics, interferometric theory via Dirac's notation, and the foundations of quantum entanglement. Duarte is a Fellow of the Australian Institute of Physics (1987) and a Fellow the Optical Society of America (1993). He has received the *Engineering Excellence Award* (1995) and the *David Richardson Medal* (2016) from the Optical Society.

T S Taylor

Travis S Taylor ('Doc' Taylor to his friends) has a doctorate in Optical Science and Engineering, a doctorate in Aerospace Systems Engineering, a master's degree in Physics and a master's degree in Aerospace Engineering, all from the University of Alabama in Huntsville. In addition, he has a master's in Astronomy from the University of Western Sydney (Australia) and a degree in Electrical

Engineering from Auburn University (Alabama). Taylor has worked on numerous programs for the Department of Defense and NASA for the past two decades. He is currently working on several advanced propulsion concepts, very large space telescopes, space-based energy systems, next generation space-launched concepts, directed energy, and nanosatellites. Highlights from his work at the U. S. Army Space and Missile Defense Command are his roles as principal investigator in the 100 Joule tunable laser program, narrow-linewidth tunable laser program, multi-wavelength oscillator-amplifier experiments, ruggedized tunable laser development, and the solid-state laser testbed experiment. He is currently focusing on high-brightness sources for quantum entanglement space communications.

He is the author of over a dozen refereed technical papers and the sole author of the textbooks *Introduction to Rocket Science and Engineering* (1st and 2nd editions), and *Introduction to Laser Science and Engineering*. In addition, over the past two decades, he has published some two dozen science fiction novels including titles such as *The Quantum Connection*, *von Neumann's War*, and *Warp Speed.*

Taylor is also an active contributor and participant in the field of science communications. As such, he has appeared in numerous television programs in the *History Channel*, *National Geographic Channel*, and *The Weather Channel*. His extracurricular activities include marathons, triathlons, CrossFit, SCUBA diving, and flying as a private pilot. He currently lives in northern Alabama with his wife and two children, just outside Huntsville, in view of the Saturn V rocket at the U. S. Space and Rocket Center.

F J Duarte and T S Taylor

Chapter 1

Introduction

The basic principles of quantum mechanics relevant to the probability amplitude of quantum entanglement, $|\psi\rangle = (|x_1, y_2\rangle - |y_1, x_2\rangle)$, are introduced. The philosophical path to quantum entanglement, Einstein, Podolsky and Rosen (EPR) → Bohm and Aharonov → Bell, and the physics path to quantum entanglement, Dirac → Wheeler → Pryce and Ward, are introduced and discussed. The introduction then concludes with a chapter by chapter description of *Quantum Entanglement Engineering and Applications* and its intended purpose.

1.1 Introduction

Quantum Entanglement Engineering and Applications is a book motivated by the desire of providing a textbook style introduction to quantum entanglement for advanced physics students, experimental physicists, scientists, and engineers. It is the opinion of the authors that the necessary background to successfully deal with the material in this book is provided by a solid background in algebra and calculus at a first year university level.

The approach is direct and particular emphasis has been given to make each chapter as transparent as possible. In this regard, the whole book is entirely based on Dirac's notation *bra ket* which is introduced in chapter 2.

At the end of each chapter a set of problems is provided. Careful attention to the material in the chapter should facilitate the solution of these problems.

The two pillars to the knowledge base exposed on this book are *The Principles of Quantum Mechanics* (Dirac 1958) and *The Feynman Lectures on Physics* (Feynman *et al* 1965). The notation and style of the book is based on the recent monograph *Fundamentals of Quantum Entanglement* (Duarte 2019).

Quantum entanglement engineering, as a general concept, has transitioned from a faint futuristic idea (Duarte 2012a, 2014) to a tangible discipline that has rapidly become the focus of considerable attention whilst even contemplating workforce requirements (Wills 2020). A new kind of wondrous engineering is in front of us and

doi:10.1088/978-0-7503-3407-5ch1

Quantum Entanglement Engineering and Applications offers a guide to open multiple doors into this entangled new world.

1.2 Essentials of quantum mechanics

From a pragmatic interferometric perspective, the very basic quantum mechanical framework needed to make the physics work is based on the following laws, or principles:

1. Theorem: The probability amplitude for a single quantum is mathematically identical to the probability amplitude for an ensemble of indistinguishable quanta.
2. The probability amplitude for a single quantum, or an ensemble of N indistinguishable quanta, is given by the Dirac–Feynman principle (Feynman *et al* 1965)

$$\langle d|s\rangle = \sum_{i=1}^{N} \langle d|\, j\rangle \langle\, j|s\rangle \tag{1.1}$$

3. The Dirac–Feynman probability amplitude is inherently linked to the *principle of superposition*: 'Any state may be considered as the result of a superposition of two or more other states ... conversely any two or more states may be superposed to give a new state' (Dirac 1958).

   In an analogy with vector notation, Feynman refers to the j states as base states… 'any quantum mechanical state can be described completely by the amplitude $\langle\, j|s\rangle$ to go into the base states ... any state in the world can be expressed as a superposition ... of base states' (Feynman *et al* 1965), see chapters 2 and 4.
4. The probability amplitudes can be represented by complex wave functions of the form (Dirac 1958)

$$\Psi = \Psi_0 e^{-i(\omega t - kx)} \tag{1.2}$$

   or more specifically

$$\langle\, j|s\rangle = \Psi(r_{j\leftarrow s})e^{-i\theta} \tag{1.3}$$

$$\langle d|\, j\rangle = \Psi(r_{d\leftarrow j})e^{-i\varphi} \tag{1.4}$$

5. The physics of quantum mechanics is *reversible*. For instance, the probability amplitude for quantum entanglement can be derived from the Dirac–Feynman principle (Duarte 2013a, 2013b)

$$\langle d|s\rangle = \sum_{j=1}^{N} \langle d|\, j\rangle \ \langle\, j|s\rangle \rightarrow |\psi\rangle = \frac{1}{\sqrt{2}}(|x\rangle_1|y\rangle_2 \pm |y\rangle_1|x\rangle_2) \tag{1.5}$$

or *vice versa* (Duarte 2019)

$$|\psi\rangle = \frac{1}{\sqrt{2}}(|x\rangle_1|y\rangle_2 \pm |y\rangle_1|x\rangle_2) \rightarrow \langle d|s\rangle = \sum_{j=1}^{N}\langle d|j\rangle\langle j|s\rangle \tag{1.6}$$

6. Additional Dirac principles as outlined by Feynman *et al* (1965)

$$\langle j|i\rangle = \delta_{ji} \tag{1.7}$$

$$\langle d|s\rangle = \langle s|d\rangle^* \tag{1.8}$$

7. Dirac's identity for indistinguishable quanta in different states (Dirac 1958)

$$|X\rangle = |a\rangle_1|b\rangle_2|c\rangle_3\ldots|g\rangle_n \tag{1.9}$$

8. Born's rule: the probability is given by multiplying the probability amplitude with its complex conjugate (Born 1926)

$$|\langle d|s\rangle|^2 = \langle d|s\rangle\langle d|s\rangle^* \tag{1.10}$$

9. The measured intensity, at a given frequency, is proportional to the probability

$$I \propto \langle d|s\rangle\langle d|s\rangle^* \tag{1.11}$$

10. 'Photons cannot be localized … and they do not behave at all like particles' (Lamb 1995).
11. Heisenberg's uncertainty principle (Heisenberg 1925)

$$\Delta p\, \Delta x \approx h \tag{1.12}$$

    For a transparent derivation of $\Delta p\, \Delta x \approx h$, from an interferometric perspective, see Duarte (2003, 2014).
12. Planck's quantum energy equation (Planck 1901)

$$E = h\nu \tag{1.13}$$

where $h = 6.62607015 \times 10^{-34}$ Js is Planck's constant. In fact, $E = h\nu$ was the birth of quantum mechanics. This birth originated in *a macroscopic experiment* (Planck 1901). This is important: **a macroscopic experiment**.

1.2.1 The quantum photon

The photon, or quantum, in quantum mechanics *is not* a particle. To use the word 'particle' in reference to the photon, in quantum entanglement, is to invite misunderstandings and to even deny the essence of quantum entanglement. In this section a series of conceptual statements that help describe the photon as the center piece of quantum entanglement are introduced:

1. 'Photons cannot be localized in any meaningful manner, and they do not behave at all like particles, whether described by a wave function or not' (Lamb 1995).
2. 'The wave function gives information about the probability of *one* photon being in a particular place and not the probable number of photons in that place … Each photon then interferes only with itself. Interference between two different photons never occurs' (Dirac 1958).
3. 'Newton thought that light was made up of particles, but then it was discovered that it behaves like a wave … We say: It is like *neither*' (Feynman *et al* 1965).

In addition to these landmark concepts the following description might also help to reinforce the non locality of the photon: 'All the indistinguishable photons illuminate the array of N slits, or grating, simultaneously. If only one photon propagates, at any given time, then that individual photon illuminates the whole array of N slits simultaneously (Duarte 2003).

As far as Dirac's dictum on interference is concerned: it should be clarified that in quantum mechanics *two indistinguishable quanta are the same quantum*, or photon. Photons emitted by two different sources that have the same frequency, $\nu_1 = \nu_2$, and the same linewidth, $\Delta\nu_1 = \Delta\nu_2$, *are the same photon.*

The two most important constants in quantum optics are the speed of light and Planck's constant: $c = 2.99792458 \times 10^8$ ms^{-1} and $h = 6.62607015 \times 10^{-34}$ Js. With the introduction of the two constants a more quantitative description of the photon can be attempted.

The photon is identified by a frequency ν, and a wavelengthλ, these quantities are related by

$$\lambda = \frac{c}{\nu} \tag{1.14}$$

The quantum energy, or the energy of the photon, as already introduced is $E = h\nu$, or

$$E = \hbar\omega \tag{1.15}$$

where $\hbar = h/2\pi$ and $\omega = 2\pi\nu$is the angular frequency. A single photon is associated with a quantum wave momentum of (de Broglie 1924)

$$p = \hbar k \tag{1.16}$$

where $k = 2\pi/\lambda$ is known as the wave number. Substitution of $2\pi/\lambda$ into equation (1.16) leads directly to

$$p = \frac{h}{\lambda} \tag{1.17}$$

The $p = \hbar k$ concept does not come into play either in quantum interference or in the interferometric derivation of the quantum entanglement probability amplitude. However, it is used in the physical approximate derivation of Heisenberg's uncertainty principle $\Delta p \, \Delta x \approx h$ (Duarte 2003, 2014) and in the heuristic derivation of Schrödinger's equation (Haken 1981; Duarte 2014).

Following Dirac (1958), a single photon is associated with the *complex wave function*

$$\psi(x, t) = \psi_0 e^{-i(\omega t - kx)} \tag{1.18}$$

Single photons can be extremely non-local and can exhibit *enormous coherence lengths*. Using Heisenberg's uncertainty principle $\Delta p \, \Delta x \approx h$, it follows that

$$\Delta x \approx \frac{c}{\Delta \nu} \tag{1.19}$$

which can be an enormous number given the very narrow linewidth $\Delta \nu$ associated with a single photon. Thus the photon can be described as an energy entity with non-local properties.

1.2.2 Planck's constant: *h*

The constant $h = 6.62607015 \times 10^{-34}$ Js is the most important constant in quantum optics and perhaps the most important constant in nature. It is crucial to quantum energy and to the uncertainty principle and its significance extends into cosmology, as illustrated by Dirac (1938, 1957).

In addition to its central significance already highlighted in $E = h\nu$ and $\Delta p \, \Delta x \approx h$, Planck's constant is central to Dirac's fine structure constant α. This was done by Dirac by defining a unit of time in units of s

$$t_1 = \left(\frac{e^2}{4\pi \varepsilon_0 m c^3} \right) \tag{1.20}$$

then he defined another time quantity

$$t_2 = \left(\frac{\hbar}{mc^2} \right) \tag{1.21}$$

and proceeded to take the ratio t_2/t_1 to arrive at

$$\alpha^{-1} = h\left(\frac{2c\varepsilon_0}{e^2} \right) \approx 137 \tag{1.22}$$

which is the dimensionless fine structure constant α (Dirac 1938). Cyclotron measurements yield $\alpha^{-1} = 137.035\ 999\ 084$ (Hanneke *et al* 2008) and α is a parameter of interest in a variety of fields including optical properties of graphene (Nair *et al* 2006), electron physics (Hanneke *et al* 2008), and quantum electrodynamics (Gabrielse *et al* 2006; Dombey 2021).

Provided that α is determined independently then the origin of h can be related to

$$h = \alpha^{-1}\left(\frac{e^2}{2c\varepsilon_0}\right) \tag{1.23}$$

However, it should be emphasized that Dirac's initial interest in α was due to cosmology. More about this in appendix A.

1.3 Ward's succinct perspectives

Ward said very little, and wrote even less, about the essence and the foundations of quantum mechanics. The next two paragraphs capture best his succinct style.

> '$(|x, y\rangle - |y, x\rangle)$ … *was my first lesson in quantum mechanics, and in a very real sense my last, since all the rest is mere technique, which can be learnt from books*' (Ward 1949).

> '*The inner mysteries of quantum mechanics require a willingness to extend one's mental processes into a strange world of phantom possibilities, endlessly branching into more and more abstruse chains of coupled logical networks, endlessly extending themselves forward and even backwards in time*' (Ward 1949).

1.4 The philosophy and the physics of quantum entanglement

As pointed out by Duarte (2012b, 2019) there are two ways to approach quantum entanglement. One is a philosophical path and the other is a physics path.

The philosophical path was initiated by Einstein *et al* (1935) in their famous EPR paper. This path is based on the premise that quantum mechanics 'does not provide a complete description of the physical reality' (Einstein *et al* 1935). That statement led to the hidden variable theories, to Bell's theorem, and to the early optical experiments on quantum entanglement.

The physical path was initiated indirectly by Dirac (1930), advanced by Wheeler (1946), and quantified by Pryce and Ward in the first paper that disclosed the superposition quantum probability based on the probability amplitude (Pryce and Ward 1947)

$$|\psi\rangle_{-}|\psi\rangle_{-}^{*} = P(\varphi_1, \varphi_2) = -\cos 2(\varphi_1 - \varphi_2) \tag{1.24}$$

of what became known as quantum entanglement.

These two approaches to quantum entanglement are described in more detail in the next two sub sections and are based on the review presented by Duarte (2019).

1.4.1 The philosophy

As already indicated, the philosophical path towards quantum entanglement was initiated by the EPR paper (Einstein *et al* 1935) and immediately followed by the

Schrödinger papers (Schrödinger 1935, 1936). Indeed, it was Schrödinger who introduced the word *entanglement* to describe the situation at hand. EPR and Schrödinger, in particular, were discontent with the non-deterministic nature of quantum mechanics. While EPR expressed the hope for a 'complete' version of quantum mechanics, Schrödinger used the word 'disconcerting' to describe his discontentment in light of the new nonlocal developments made possible by quantum mechanics.

The philosophical path can be briefly summarized as: doubts on the completeness of quantum mechanics → hidden variable theories → Bell's theorem → Aspect's experiments. A more nuance description of this path is given in figure 1.1. In this figure the direct path of citations is indicated by the solid arrows downward.

Important publications that outline this path are: Einstein *et al* (1935), Schrödinger (1935, 1936), Bohm and Aharonov (1957), Bell (1964), Clauser *et al* (1969), Aspect *et al* (1981, 1982a, 1982b).

Important footnote: the concept of incompatibility between quantum mechanics and hidden variables was first introduced in 1932 by von Neumann, however, it was claimed that the proof leading to the von Neumann (1932) conclusion was found 'wanting' (Bell 1964). This subject is discussed in detail in chapters 8 and 10. Please refer to chapter 10 for further discussions on the philosophical aspects of quantum entanglement.

1.4.2 The physics

The physical path toward quantum entanglement was initiated in a complete vacuum of philosophical arguments by a paper of Dirac on pair theory (Dirac 1930). In the

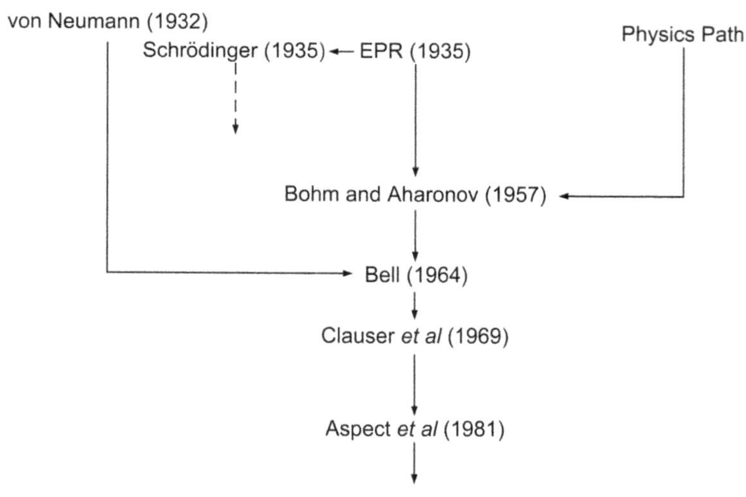

Figure 1.1. The philosophical path to quantum entanglement. The papers by Schrödinger received intermittent citations for many years and only began to be cited more regularly since the 1990s. The assigned years refer to literature emergence dates. General comments about figures 1.1–1.3: solid arrowed lines indicate a direct line of citation and the main citation path runs vertically downwards (from Duarte 2019).

pair theory Dirac predicted that a slow positron beam incident on an electron at rest created a pair of photons, or a pair of quanta (Dombey 2021)

$$e^{+} + e^{-} \rightarrow \gamma_1 + \gamma_2 \tag{1.25}$$

which is also expressed in the literature as (Dalitz and Duarte 2000)

$$e^{+}e^{-} \rightarrow \gamma_1\gamma_2 \tag{1.26}$$

Dirac's pair theory predicted that the polarization of these two gamma quanta would be polarized orthogonal to each other. Inspired by Dirac's paper Wheeler sought to envision an experiment to test the pair theory and wrote: '*if one of these photons is linearly polarized in one plane, then the photon that goes off in the opposite direction with equal momentum is linearly polarized in the perpendicular plane*' (Wheeler 1946). This statement by Wheeler captures the very essence of quantum entanglement and remains until today perhaps the best and most transparent description of what quantum entanglement is.

Almost immediately following the paper by Wheeler, Pryce and Ward published a cryptic ~½ page paper entitled *Angular correlation effects with annihilation radiation* (Pryce and Ward 1947). This paper was seismic in two aspects: first, it provided the very first diagram on a quantum entanglement experiment and, second, it calculated the scattering cross section of the annihilation process. This equation, which has the form of

$$\sigma = A - B\cos 2(\varphi_1 - \varphi_2) \tag{1.27}$$

is a function of the quantum superposition probability for quantum entanglement

$$P(\varphi_1, \varphi_2) = -\cos 2(\varphi_1 - \varphi_2) \tag{1.28}$$

calculated from the normalized version of the probability amplitude for quantum entanglement (Ward 1949)

$$|\psi\rangle = (|x_1, y_2\rangle - |y_1, x_2\rangle) \tag{1.29}$$

Within one year quantum entanglement experiments via electron–positron annihilation were conducted with mild success comparing the theoretical predictions of Pryce and Ward with experimental measurements (Hanna 1948; Bleuler and Bradt 1948). A more refined version of the annihilation experiments was conducted by Wu and Shaknov (1950) yielding better agreement, within 4%, between theory and experiment.

The physical path toward quantum entanglement can be summarized as: Dirac's pair theory→Wheeler's paper→Pryce and Ward theory→annihilation experiments. A more nuance description of this path is given in figure 1.2. In this figure the direct path of citations is indicated by the solid arrows downward.

In chronological order, notable publications outlining the physical path toward quantum entanglement include: Dirac (1930), Wheeler (1946), Pryce and Ward (1947), Hanna (1948), Bleuler and Bradt (1948), Wu and Shaknov (1950), Kasday *et al* (1975).

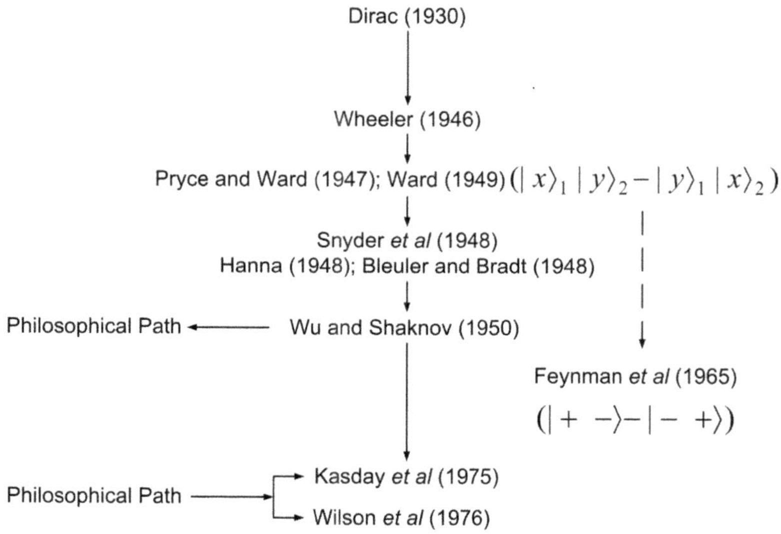

Figure 1.2. The physics path to quantum entanglement (from Duarte 2019).

Some important footnotes are important here: in 1948 a paper by Snyder *et al* (1948) utilized *wave functions* to express the probability amplitude of quantum entanglement in a form similar to equation (1.29) thus independently confirming the Pryce and Ward physics. However, the sign in this equation was positive thus introducing a discrepancy with the available experimental results (Dombey 2021).

A second footnote concerns the work of Feynman in the subsequent derivation of the probability amplitude for quantum entanglement: while discussing the hyperfine splitting of hydrogen, without explanation, he allows whole states to be described by notation like $|++\rangle$, $|+-\rangle$, $|-+\rangle$, and $|--\rangle$, just as Ward did in his thesis (Ward 1949), which lead to probability amplitudes of the form (Feynman *et al* 1965)

$$|IV\rangle = \frac{1}{\sqrt{2}}(|+\ -\rangle - |-\ +\rangle) \tag{1.30}$$

Apparently, it is this equation that many authors use in the literature today without referencing its origin. Feynman's contribution and approach to quantum entanglement is discussed in detail in chapters 6 and 10.

1.5 Quantum entanglement as a discipline

It is interesting to note that the iconic word introduced to the field by Schrödinger, *entanglement*, appears to have been linked directly with Pryce and Ward, and Aspect, type quantum entanglement experiments only in the mid to late 1980s, see for example, Ghirardi *et al* (1987).

Since its controversial philosophical beginnings, or its stealth physics beginnings, the field of quantum entanglement has grown enormously. The cumulative publication number on quantum entanglement is extraordinarily large. Hence, it is impossible to provide an accurate description of the whole field other than a very

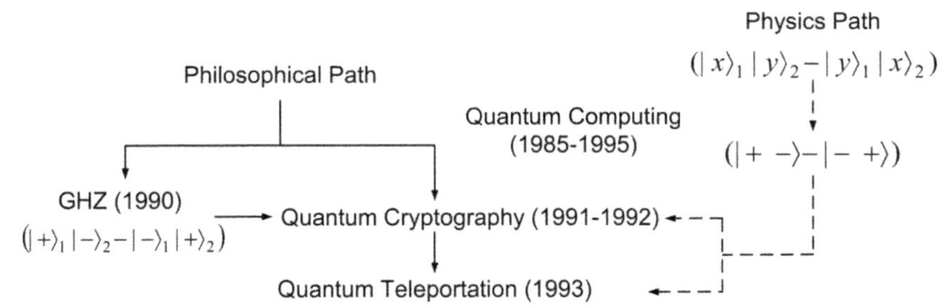

Figure 1.3. Approximate overview of the field of quantum entanglement. The philosophical path converges directly into quantum cryptography and quantum teleportation. The influence of the physical path is limited to the widespread use of $|\psi\rangle = (|x\rangle_1|y\rangle_2 - |y\rangle_1|x\rangle_2)$ in its various forms, see text (from Duarte 2019).

high level overview of its main features. In this regard, there are four main branches: quantum computing, quantum cryptographic communications, quantum entanglement metrology, and quantum teleportation. An overview of the field is given in figure 1.3.

Some relevant observations on the information conveyed in figure 1.3 are the following: quantum cryptography is positioned chronologically as having started in 1991 since that corresponds to the introduction of entangled photon pairs to the subject (Ekert 1991). However, the conceptual bases of quantum cryptography, were considered earlier by Wiesner (1983) and Bennett and Brassard (1984). Quantum teleportation was introduced by Bennett *et al* (1993).

The field of quantum computing began independently of quantum entanglement discussions. Some of the first papers on the subject were introduced by Feynman (1985, 1986). It is worth mentioning that, as typical of Feynman's style, his approach to quantum computing was his own and thus it did not follow the established philosophical path.

The term *qbit*, short for quantum bit, was introduced by Schumacher (1995) as qubit. Subsequently, Steane (1998) dealt directly with EPR–Bell notions. The applications of quantum entanglement are discussed in chapter 8.

1.6 Quantum entanglement engineering and applications

In this section an overview of the content of each chapter in *Quantum Entanglement Engineering and Applications* is provided:

Chapter 1: *Introduction*: The basic principles of quantum mechanics relevant to the probability amplitude of quantum entanglement, $|\psi\rangle = (|x_1, y_2\rangle - |y_1, x_2\rangle)$, are introduced. The philosophical path to quantum entanglement, EPR → Bohm and Aharonov →Bell, and the physics path to quantum entanglement, Dirac → Wheeler → Pryce and Ward, are introduced and discussed. The introduction then concludes with a chapter-by-chapter description of *Quantum Entanglemen Engineering and Applications* and its intended purpose.

Chapter 2: *Dirac's notation for quantum entanglement*: An introduction to Dirac's notation is given. The exposition begins at a fundamental level using experimental configurations applicable to generalized *N*-slit interferometry. Once the basics are explained the treatment advances to apply Dirac's notation to quantum entanglement situations from the usual $n = N = 2$ to generalized configurations such as $n = N = 2^1, 2^2, 2^3, 2^4 \ldots 2^r$ in addition to the cases of $n = N = 3$and $n = N = 6$. It is shown that all these configurations can be described by the generalized equation for the probability amplitudes of quantum entanglement. Dirac's identities crucial to quantum entanglement are also introduced.

Chapter 3: *Indistinguishability*: The central role that the concept of indistinguishability plays in quantum interference, quantum entanglement, and quantum mechanics as a whole is discussed.

Chapter 4: *Quantum interferometry via Dirac's bra ket notation*: The probability amplitudes applicable to the N-slit interferometer and various beam-splitter based interferometers, such as the Mach–Zehnder interferometer, the Michelson interferometer, and the Sagnac interferometer are derived. The attention then focuses on the derivation of the matrix for the Hadamard gate and on interferometers based on a single beam splitter.

Chapter 5: *Vectors, matrices, and tensors for quantum entanglement*: Vector products, matrices, and tensor elements applicable to quantum entanglement are reviewed in a direct and terse style. This review includes the vector and tensor products relevant to quantum entanglement probability amplitudes. Matrix algebra techniques directly applicable to quantum optics are described. Pauli matrices are utilized to illustrate the mathematics of eigenvalues and eigenvectors. Matrix unitary properties are also described.

Chapter 6: *Five avenues to the probability amplitude of quantum entanglement*: The five paths to quantum entanglement are discussed quantitatively. Starting with Ward's heuristic derivation, the chapter continues to describe Feynman's two-state approach, the N-slit interferometric approach, the Pauli matrix approach, and the Hadarmard matrix approach. The chapter concludes with a discussion on the fundamentality of quantum entanglement versus the fundamentality of quantum interference at the foundations of quantum mechanics.

Chapter 7: *Quantum entanglement in matrix notation*: The probability amplitudes for quantum entanglement for $n = N = 2$ are expressed in 2×2 matrix form. The explicit mathematical nexus between the probability amplitudes for quantum entanglement with the Pauli matrices and the Hadamard matrix is described. From this approach, a series of new 2×2 optical matrices are introduced.

Chapter 8: *Quantum entanglement applications*: The application of the probability amplitude for quantum entanglement, $|\psi\rangle = (|x\rangle_1|y\rangle_2 \pm |y\rangle_1|x\rangle_2)$, to quantum cryptography communications, quantum teleportation, quantum computing, and quantum metrology is described. An overview of additional applications is also provided.

Chapter 9: *Satellite networks for quantum communications*: Parameters relevant to satellite communications are reviewed as well as existing data from satellite quantum entanglement experiments. Various satellite network concepts, of practical relevance, are considered.

Chapter 10: *Quantum entanglement and the interpretation of quantum mechanics*: The subject of interpretations of quantum mechanics is examined and discussed from a pragmatic perspective. The thoughts of Born, Dirac, Dyson, Feynman,

Lamb, and Ward are applied to further develop a direct pragmatic paradox-free perspective of quantum mechanics. Particular attention is given to Dyson's *first layer* and *second layer* description of quantum mechanics. This perspective enables the description of quantum interferometry measurements and quantum entanglement measurements in a cohesive indeterministic approach via the concept of *superposition probabilities.*

Appendix A: *More on Dirac's notation: applications to laser cavities and interferometry*: The use of Dirac's quantum notation in the description of macroscopic highly coherent high-power tunable laser sources is described. These concepts are further applied to the description of sub-microscopic electrically-pumped coherent organic semiconductor emitters. The transition from quantum to classical interference is also described.

An effort has been made to make each chapter, as much as possible, self-contained thus minimizing the need to depend on material presented in other chapters. This means that, as a matter of presentation style, necessary equations and concepts are explicitly re-stated whenever there is a pedagogical need to do so: *repetitio est mater studiorum.*

1.7 Intent

Advances in physics have been made by physicists rather than by philosophers. Thus the path chosen by *Quantum Entanglement Engineering and Applications* towards quantum entanglement is the physical path rather than the philosophical alternative. This saves a lot of time and energy and neutralizes unnecessary arguments and discussions.

As hinted by John Ward, quantum mechanics is a field that can utilize a lot of book work and mathematical techniques. In *Quantum Entanglement Engineering and Applications* (QEEA) the focus is on the essentials of Dirac's notation that make things work. In this regard, QEEA has been written specifically for engineers, physicists, scientists, and practically minded individuals in general, who are interested in understanding the necessary essentials so as to make experiments and technology work. From an optimistic perspective, it is the hope of the authors, that this approach will inspire some bright minds to extend the field of quantum entanglement and its technology. After all, Dirac himself was an engineer before he became a physicist.

One final observation: albeit the word engineering is part of the title of QEEA the emphasis is on the physics that facilitates the engineering rather than on present day technological elements and systems that might quickly become superseded by new developments in the technology. Also, even though QEEA refers specifically and entirely to space-to-space configurations, the principles taught apply to quantum entanglement configurations utilizing other propagation means such as fibers.

Problems

1.1 Expand the probability amplitude given in equation (1.1) for $N = 2$.

1.2 Write an expression for the conjugate of $\langle j|s\rangle = \Psi(r_{j\leftarrow s})e^{-i\theta}$.

1.3 Write an expression for the conjugate of $\psi(x, t) = \psi_0 e^{-i(\omega t - kx)}$.
1.4 Show that $\Delta p \, \Delta x \approx h$ can be expressed as $\Delta \nu \Delta x \approx c$.
1.5 Show that $\Delta p \, \Delta x \approx h$ can be expressed as $\Delta \nu \Delta t \approx 1$.
1.6 For $\Delta \nu \approx 100$ Hz calculate the coherence length Δx.
1.7 Show that the probability amplitudes given by equations (1.29) and (1.30) are equivalent. Note: background to do this problem is given in chapter 7.

References

Aspect A, Grangier P and Roger G 1981 Experimental tests of realistic local theories via Bell's theorem *Phys. Rev. Lett.* **47** 460–3

Aspect A, Grangier P and Roger G 1982a Experimental realization of Einstein–Podolsky–Rosen–Bohm gedanken experiment: a new violation of Bell's inequalities *Phys. Rev. Lett.* **49** 91–4

Aspect A, Grangier P and Roger G 1982b Experimental test of Bell's inequality using time-varying analyzers *Phys. Rev. Lett.* **49** 1804–7

Bell J S 1964 On the Einstein–Podolsky–Rosen paradox *Physics* **1** 195–200

Bennett C H and Brassard G 1984 Quantum cryptography: public key distribution and coin tossing *Proc. IEEE Int. Conf. Comput. Syst. Signal Process* (Bangalore: IEEE)

Bennett C H, Brassard G, Crépeau C, Jozsa R, Peres A and Wootters W K 1993 Teleporting an unknown quantum state via dual classical and Einstein–Podolsky–Rosen channels *Phys. Rev. Lett.* **70** 1895–9

Bleuler E and Bradt H L 1948 Correlation between the states of polarization of the two quanta of annihilation radiation *Phys. Rev.* **73** 1398

Bohm D and Aharonov Y 1957 Discussion of experimental proof for the paradox of Einstein, Rosen, and Podolsky *Phys. Rev.* **108** 1070–6

Born M 1926 Zur quantenmechanik der stoßvorgänge *Z. Phys.* **37** 863–27

Born M, Heisenberg W and Jordan P 1926 Zur quantenmechanik II *Z. Phys.* **35** 557–615

Clauser J F, Horner M A, Shimony A and Holt R A 1969 Proposed experiment to test local hidden variable theories *Phys. Rev. Lett.* **23** 880–3

Dalitz R H and Duarte F J 2000 John Clive Ward *Phys. Today* **53** 99–100

de Broglie L 1924 A tentative theory of light quanta *London Edinburgh Dublin Philos. Mag. J. Sci.* **47** 446–58

Dicke R H 1957 Gravitation without a principle of equivalence *Rev. Mod. Phys.* **29** 363–76

Dirac P A M 1930 On the annihilation of electrons and protons *Math. Proc. Camb. Phil. Soc.* **2** 361–75

Dirac P A M 1938 A new basis for cosmology *Proc. R. Soc.* **A165** 199–208

Dirac P A M 1958 *The Principles of Quantum Mechanics* 4th edn (Oxford: Oxford University Press)

Dombey N 2021 John Clive Ward *Biogr. Mems. Fell. R. Soc* **70** 419–40

Duarte F J 2003 *Tunable Laser Optics* (New York: Elsiever Academic)

Duarte F J 2012a *Laser Physicist* (New York: Optics Journal)

Duarte F J 2012b The origin of quantum entanglement experiments based on polarization measurements *Eur. Phys. J.* H **37** 311–8

Duarte F J 2013a The probability amplitude for entangled polarizations: an interferometric approach *J. Mod. Opt.* **60** 1585–7

Duarte F J 2013b Tunable laser optics: applications to optics and quantum optics *Prog. Quantum Electron.* **37** 326–47

Duarte F J 2014 *Quantum Optics for Engineers* (New York: CRC Press)

Duarte F J 2019 *Fundamentals of Quantum Entanglement* (Bristol: IOP Publishing)

Einstein A, Podolsky B and Rosen N 1935 Can quantum mechanical description of physical reality be considered complete? *Phys. Rev.* **47** 777–80

Ekert A K 1991 Quantum cryptography based on Bell's theorem *Phys. Rev. Lett.* **67** 661–3

Feynman R P 1985 Quantum mechanical computers *Opt. News* **11** 11–20

Feynman R P 1986 Quantum mechanical computers *Found. Phys.* **16** 507–31

Feynman R P, Leighton R B and Sands M 1965 *The Feynman Lectures on Physics* vol III (Reading, MA: Addison-Wesley)

Gabrielse G, Hanneke D, Kinoshita T, Nio M and Odom B 2006 New determination of the fine structure constant from the electron g value in QED *Phys. Rev. Lett.* **97** 030802

Ghirardi G C, Rimini A and Weber T 1987 Disentanglement of quantum functions: Answer to comment on 'Unified dynamics for microscopic and macroscopic systems' *Phys. Rev.* D **36** 3287–789

Hanna R C 1948 Polarization of annihilation radiation *Nature* **162** 332

Hanneke D, Fogwell S and Gabrielse G 2008 New measurement of the electron magnetic moment and the fine structure constant *Phys. Rev. Lett.* **100** 120801

Haken H 1981 *Light* (Amsterdam: North Holland)

Heisenberg W 1925 Uber quantenthoretische umdeutung kinematischer und mechanischer beziehungen *Z. Phys.* **33** 879–93

Kasday L R, Ullman J D and Wu C S 1975 Angular correlation of Compton-Scattered annihilation photons and hidden variables *Il Nuovo Cimento* **25** 633–61

Lamb W E 1995 Anti-photon *Appl. Phys.* B **60** 77–84

Nair R R, Blake P, Grigorenko A N, Novoselov K S, Booth T J, Stauber T, Peres N M R and Geim A K 2006 Fine structure constant defines visual transparency of graphene *Science* **320** 1308

Planck M 1901 Ueber das gesetz der energieverteilung im normalspectrum *Ann. Phys.* **309** 553–63

Pryce M L H and Ward J C 1947 Angular correlation effects with annihilation radiation *Nature* **160** 435

Schrödinger E 1935 Discussion of probability relations between separated systems *Math. Proc. Camb. Phil. Soc.* **31** 555–63

Schrödinger E 1936 Probability relations between separated systems *Math. Proc. Camb. Phil. Soc.* **32** 446–52

Schumacher B 1995 Quantum coding *Phys. Rev. Lett.* **51** 2738–47

Snyder H S, Pasternack S and Hornbostel J 1948 Angular correlations of scattered annihilated radiation *Phys. Rev.* **73** 440–8

Steane A 1998 Quantum computing *Rep. Prog. Phys.* **61** 117–74

von Neumann J 1932 *Mathematische Grundlagen der Quanten-Mechanik* (Berlin: Springer)

Ward J C 1949 *Some Properties of the Elementary Particles* (Oxford: Oxford University Press)

Wiesner S 1983 Conjugate coding *ACM SIGACT News* **15** 78–88

Wheeler J A 1946 Polyelectrons *Ann. NY Acad. Sci.* **48** 219–38

Wills S 2020 A work force for quantum tech *Opt. Photon. News* **31** 15

Wu C S and Shaknov I 1950 The angular correlation of scattered annihilation radiation *Phys. Rev.* **77** 136

IOP Publishing

Quantum Entanglement Engineering and Applications

F J Duarte and T S Taylor

Chapter 2

Dirac's notation for quantum entanglement

An introduction to Dirac's notation is given. The exposition begins at a fundamental level using experimental configurations applicable to generalized N-slit interferometry. Once the basics are explained the treatment advances to apply Dirac's notation to quantum entanglement situations from the usual $n = N = 2$ to generalized configurations such as $n = N = 2^1, 2^2, 2^3, 2^4 \ldots 2^r$ in addition to the cases of $n = N = 3$ and $n = N = 6$. It is shown that all these configurations can be described by the generalized equation for the probability amplitudes of quantum entanglement. Dirac's identities crucial to quantum entanglement are also introduced.

2.1 Introduction

In this chapter the Dirac–Feynman principle, which is inherently applicable to N-slit interferometric configurations, is utilized to illustrate the fundamentals of Dirac's quantum notation. Beyond a detailed description of probability amplitudes and superposition probabilities applicable to quantum N-slit interferometry the interferometric framework is extended to derive, in conjunction with a crucial Dirac identity, the equations for quantum entanglement. At first this is done for the case of a quanta pair $n = 2$ and two propagation channels $N = 2$. The treatment is then methodically extended to the $n = N = 4$, $n = N = 8$, $n = N = 16$, $n = N = 2^1, 2^2, 2^3, 2^4 \ldots 2^r$ to finally culminate with generalized quantum entanglement equations. Attention is then turned to the special cases of $n = N = 3$, and $n = N = 6$, which involve Hamilton's quaternions.

This chapter is a review that integrates various aspects previously published in separate publications (Duarte 2014, 2015, 2018, 2019, Duarte and Taylor 2017). In this regard, this review offers a unified treatment on the subject. For consistency the notation and style previously used in Duarte (2014, 2019) is preserved.

2.2 Dirac's *bra ket* notation

Dirac disclosed this visionary probabilistic notation in a paper entitled *A new notation for quantum mechanics* (Dirac 1939). In this notation a *bra* vector is denoted by $\langle|$ and a *ket* vector is denoted by $|\rangle$. In Dirac's notation, the *probability amplitude* to go from the emission s plane to the detection d plane is described by the *bra ket* $\langle d|s\rangle$ which is a *complex number*. It is important to indicate that the d plane is also the interferometric plane. Note that the propagation from the s plane to the d plane is expressed *in reverse* by $\langle d|s\rangle$.

The propagation of a single photon from plane s to plane d via the passage through an intermediate plane j is described by the probability amplitude (Feynman *et al* 1965)

$$\langle d|s\rangle = \langle d|j\rangle\langle j|s\rangle \tag{2.1}$$

Replacing the single intermediate plane by a transmission diffraction grating offering passage alternatives from 1 to N, that is $j = 1, 2, 3 \ldots, N$, leads to the configuration of an N-slit interferometer (see figure 2.1) and the corresponding superposition probability amplitude becomes

$$\langle d|s\rangle = \sum_{j=1}^{N}\langle d|j\rangle\langle j|s\rangle \tag{2.2}$$

This equation is intrinsic to N-slit interferometry and the sum over N emphasizes that in quantum mechanics *all* propagation paths must be considered (Feynman and Hibbs 1965). Beyond its usefulness to interferometry, in general this equation is a profound statement in statistical physics and it is referred to as the *Dirac–Feynman interferometric principle* in this book.

In figure 2.1 the distance from the plane s to j is denoted as $D_{\langle j|s\rangle}$ while the intra interferometric distance from plane j to d is denoted as $D_{\langle d|j\rangle}$.

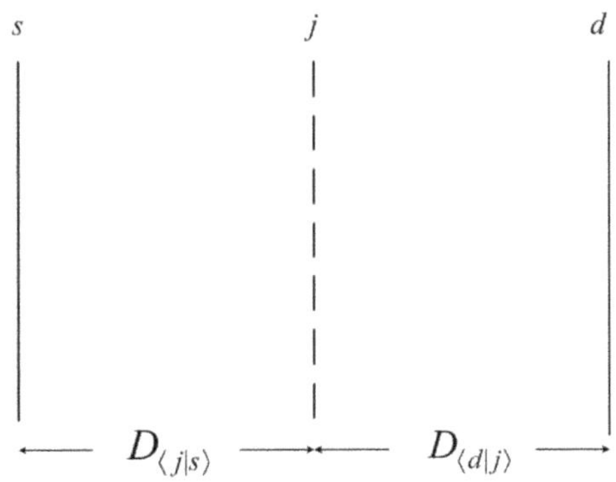

Figure 2.1. Schematic describing the optics for single-photon propagation from plane s to d, via the transmission interferometric plane j. Plane j is comprised of an array of N-slits. Interference takes place between plane j and the interferometric, or detection, plane d. The intra interferometric distance from plane j to d is denoted as $D_{\langle d|j\rangle}$. This schematics is also applicable to illumination by a population of indistinguishable photons (Duarte 2019).

Addition of an intermediate plane k, between j and d, as depicted in figure 2.2, leads to the superposition probability amplitude

$$\langle d|s\rangle = \sum_{k=1}^{N}\sum_{j=1}^{N}\langle d|k\rangle\langle k|j\rangle\langle j|s\rangle \tag{2.3}$$

in this configuration the intra interferometric distances are $D_{\langle j|s\rangle}$, $D_{\langle k|j\rangle}$ and $D_{\langle d|k\rangle}$.

Introducing two extra N-slit intra interferometric planes, l and m, leads to the superposition probability amplitude

$$\langle d|s\rangle = \sum_{m=1}^{N}\sum_{l=1}^{N}\sum_{k=1}^{N}\sum_{j=1}^{N}\langle d|m\rangle\langle m|l\rangle\langle l|k\rangle\langle k|j\rangle\langle j|s\rangle \tag{2.4}$$

Here, see figure 2.3, the intra interferometric distances are denoted by $D_{\langle j|s\rangle}$, $D_{\langle k|j\rangle}$, $D_{\langle l|k\rangle}$, $D_{\langle m|l\rangle}$, and $D_{\langle d|m\rangle}$. It should be noted that in the probability amplitudes expressed in equations (2.2)–(2.4) the series of states on the RHS are the *base states* while the singular state on the LHS, in each case, is a *superposition state*. For further discussions on the Dirac–Feynman principle see chapter 10.

2.3 Dirac's notation in *N*-slit interferometers

Photons are non-local, they 'cannot be localized' (Lamb 1995). 'All the indistinguishable photons illuminate the array of N-slits, or grating, simultaneously. If only one photon propagates, at any given time, then the individual photon illuminates the whole array of N slits simultaneously' (Duarte 2003). What this means is that when a photon encounters a beam splitter … the same photon goes both ways.

Dirac refers directly to practical macroscopic interferometry while utilizing quantum principles (Dirac 1958). He refers to a monochromatic beam composed of a large number of photons. Photons are split and then they are made to interfere. It is under

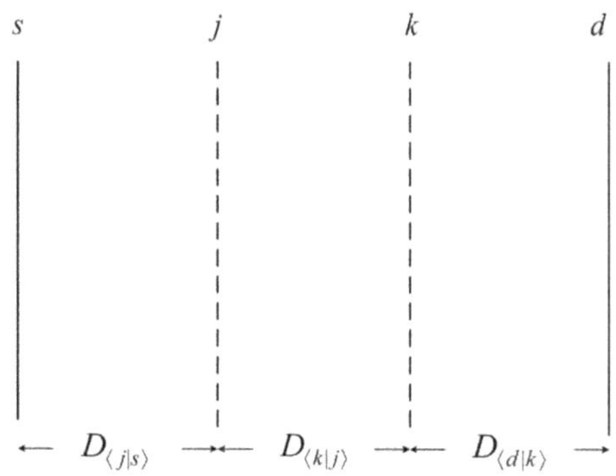

Figure 2.2. Schematic describing the optics for single-photon propagation from plane s to d, via the transmission interferometric planes j and k. Planes j and k are configured by N-slits arrays. *Measurable interference* takes place between plane k and the interferometric, or detection, plane d. The intra interferometric distances are denoted by $D_{\langle j|s\rangle}$, $D_{\langle k|j\rangle}$ and $D_{\langle d|k\rangle}$. This schematics is also applicable to illumination by a population of indistinguishable photons (Duarte 2019).

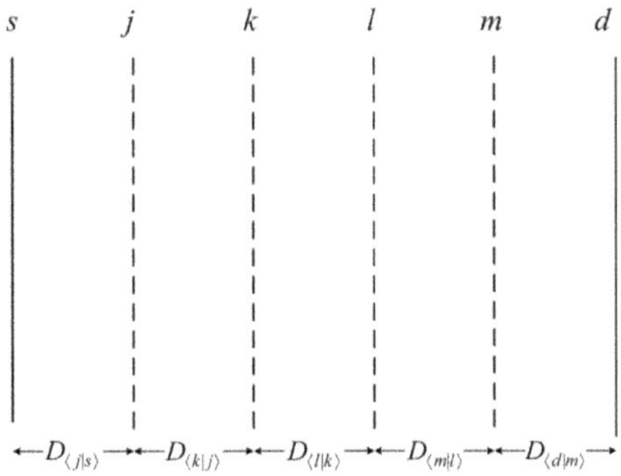

Figure 2.3. Schematic describing the optics for single-photon propagation from plane s to d, via the transmission interferometric planes j, k, l, and m. Planes j, k, l, and m are configured by N-slits arrays. *Measurable interference* takes place between plane m and the interferometric, or detection, plane d. The intra interferometric distances are denoted by $D_{\langle j|s\rangle}$, $D_{\langle k|j\rangle}$, $D_{\langle l|k\rangle}$, $D_{\langle m|l\rangle}$, and $D_{\langle d|m\rangle}$. This schematics also applicable to illumination by a population of indistinguishable photons.

these premises that he advanced the concept that 'each photon then interferes only with itself' (Dirac 1958). His description is perfectly applicable to an ensemble of indistinguishable photons as would be the case of N-slit interference using a single-transverse-mode single-longitudinal-mode narrow-linewidth laser (Duarte 1999, 2003).

Probability amplitudes accurately describe the physics but they are purely mathematical entities. To transition from a probability amplitude to a *probability* then $\langle d|s\rangle$ must be multiplied with its complex conjugate $\langle d|s\rangle^*$, according to Born's rule (Born 1926), that is

$$\langle d|s\rangle\langle d|s\rangle^* = \left(\sum_{j=1}^{N}\langle d|j\rangle\langle j|s\rangle\right)\left(\sum_{j=1}^{N}\langle d|j\rangle\langle j|s\rangle\right)^* \tag{2.5}$$

Following Dirac, the individual probability amplitudes can be represented by *wave functions of ordinary wave optics* (Dirac 1958). Thus the individual probability amplitudes are complex functions expressed as

$$\langle j|s\rangle = \Psi(r_{j,s})e^{-i\theta_j} \tag{2.6}$$

$$\langle d|j\rangle = \Psi(r_{d,j})e^{-i\phi_j} \tag{2.7}$$

where, θ_j and ϕ_j are the phase terms associated with the incidence and diffraction waves, respectively. Using these probability amplitudes, the Dirac–Feynman probability amplitude can now be expressed in its complex wave-function form as

$$\langle d|s\rangle = \sum_{j=1}^{N}\Psi(r_j)e^{-i\Omega_j} \tag{2.8}$$

where

$$\Psi(r_j) = \Psi(r_{d,j})\Psi(r_{j,s}) \tag{2.9}$$

and

$$\Omega_j = (\theta_j + \phi_j) \tag{2.10}$$

Next, to reveal the probability, Born's rule (Born 1926) is applied by multiplying equation (2.8) with its complex conjugate

$$|\langle d|s\rangle|^2 = \langle d|s\rangle\langle d|s\rangle^* \tag{2.11}$$

leading to the generalized interferometric probability equation in complex form

$$\langle d|s\rangle\langle d|s\rangle^* = \sum_{j=1}^{N}\Psi(r_j)\sum_{m=1}^{N}\Psi(r_m)e^{i(\Omega_m-\Omega_j)} \tag{2.12}$$

Expanding the complex probability equation (2.12) and using the geometrical identity

$$2\cos(\Omega_m - \Omega_j) = e^{-i(\Omega_m-\Omega_j)} + e^{i(\Omega_m-\Omega_j)} \tag{2.13}$$

renders the explicit generalized interferometric probability equation in one dimension (Duarte and Paine 1989, Duarte 1991, 1993)

$$\langle d|s\rangle\langle d|s\rangle^* = \sum_{j=1}^{N}\Psi(r_j)^2 + 2\sum_{j=1}^{N}\Psi(r_j)\left(\sum_{m=j+1}^{N}\Psi(r_m)\cos(\Omega_m - \Omega_j)\right) \tag{2.14}$$

The interferometric equations (2.12) and (2.14) are completely equivalent, albeit one is in complex form and the other is completely real. Here it should be emphasized again that these interferometric equations are *probabilities* that can lead, as explained in section 2.6, to measurable intensities.

The interferometric superposition probability equation (2.5) is applicable to single-photon propagation. The interferometric probability equations (2.12) and (2.14) though applicable to single-photon propagation are also applicable to the propagation of *ensembles of indistinguishable photons* (Duarte 1993, 2003). Ensembles, or populations, of indistinguishable photons are available from highly coherent sources such as single-transverse-mode single-longitudinal-mode narrow-linewidth lasers.

The correct quantum mechanical approach demands that the whole superposition probability $\langle d|s\rangle\langle d|s\rangle^*$, which is based in the whole superposition probability amplitude $\langle d|s\rangle$, be considered in the calculations in order to render agreement with measurements. No individual component state of the superposition can be ignored or neglected. Equation (2.14) has been experimentally confirmed in the $2 \leqslant N \leqslant 2000$ range and it has also been successfully applied to calculate diffraction via single slits modeled as being comprised of arrays of hundreds of imaginary sub-slits (Duarte 1993).

In this interferometric trinity, equation (2.14) is sometimes confused with the classical equation for N-slit interference. This topic is discussed further, and clarified, in appendix A.

The generalized interferometric probability equation can also be expressed in two dimensions (Duarte 1995, 2014, 2015),

$$\langle d|s\rangle\langle d|s\rangle^* = \sum_{z=1}^{N}\sum_{y=1}^{N}\Psi(r_{zy})\sum_{q=1}^{N}\sum_{p=1}^{N}\Psi(r_{pq})e^{i(\Omega_{qp}-\Omega_{zy})} \tag{2.15}$$

and in three dimensions (Duarte 1995, 2014, 2015),

$$\langle d|s\rangle\langle d|s\rangle^* = \sum_{z=1}^{N}\sum_{y=1}^{N}\sum_{x=1}^{N}\Psi(r_{zyx})\sum_{q=1}^{N}\sum_{p=1}^{N}\sum_{r=1}^{N}\Psi(r_{qpr})e^{i(\Omega_{qpr}-\Omega_{zyx})} \tag{2.16}$$

The quantum probabilities given in this chapter, that is equations (2.12), (2.14), (2.15), and (2.16), can either be used to predict measured interferometric distributions or to reproduce measured interferometric distributions (Duarte 1993, 1995).

Also, it should be noted that the N-slit interferometric probability described in equations (2.12) and (2.14) has been found to be consistent with equations developed to describe the interference of electrons developed from the principles of vacuum quantum fluctuation (Madrid et al 2020).

2.4 Semi coherent interference

The explicit generalized interferometric probability equation in one dimension can be extended in scope to the domain of semi coherence where an array of wavelengths, λ_1, λ_2, $\lambda_3 \ldots \lambda_n$, illuminate the diffraction grating. Under those circumstances the generalized interferometric probability equation becomes (Duarte 2014, 2015)

$$\sum_{\lambda=\lambda_1}^{\lambda_n}|\langle d|s\rangle|_\lambda^2 = \sum_{\lambda=\lambda_1}^{\lambda_n}\left(\sum_{j=1}^{N}\Psi(r_j)_\lambda\sum_{m=1}^{N}\Psi(r_m)_\lambda e^{i(\Omega_m-\Omega_j)}\right) \tag{2.17}$$

The physical meaning of this equation is the following: for a given quantum of a define wavelength, λ_1 lets say, the interferometric probability equation is

$$|\langle d|s\rangle|_{\lambda_1}^2 = \left(\sum_{j=1}^{N}\Psi(r_j)_{\lambda_1}\sum_{m=1}^{N}\Psi(r_m)_{\lambda_1}e^{i(\Omega_m-\Omega_j)}\right) \tag{2.18}$$

for λ_2 it is

$$|\langle d|s\rangle|_{\lambda_2}^2 = \left(\sum_{j=1}^{N}\Psi(r_j)_{\lambda_2}\sum_{m=1}^{N}\Psi(r_m)_{\lambda_2}e^{i(\Omega_m-\Omega_j)}\right) \tag{2.19}$$

and so on. The overall measurement is a cumulative interferometric pattern corresponding to λ_1, λ_2, $\lambda_3 \ldots \lambda_n$ and the interferometric probability is as given in equation (2.17). The detector registers the cumulative broader interferometric pattern described by equation (2.17) with the sharp interferometric profiles associated with quanta of a single frequency, described by equations (2.18) and (2.19), contributing to the overall cumulative interferogram (Duarte 2015).

2.5 Expanded series of *N*-slit quantum interference probabilities

For experimentalists and engineers it is useful to express the generalized interferometric probability equations in an explicit expanded form. For this, starting from

$$\langle d|s\rangle\langle d|s\rangle^* = \sum_{j=1}^{N}\Psi(r_j)^2 + 2\sum_{j=1}^{N}\Psi(r_j)\left(\sum_{m=j+1}^{N}\Psi(r_m)\cos(\Omega_m - \Omega_j)\right) \tag{2.20}$$

it can be shown that for the $N = 2$, or double-slit interferometer, the superposition probability becomes (Duarte 1993)

$$|\langle d|s\rangle|^2 = \Psi(r_1)^2 + \Psi(r_2)^2 + 2\Psi(r_1)\Psi(r_2)\cos(\Omega_2 - \Omega_1) \tag{2.21}$$

for the $N = 3$, or triple slit interferometer the superposition probability becomes (Duarte 2014)

$$\begin{aligned} |\langle d|s\rangle|^2 &= \Psi(r_1)^2 + \Psi(r_2)^2 + \Psi(r_3)^2 \\ &+ 2(\Psi(r_1)\Psi(r_2)\cos(\Omega_2 - \Omega_1) + \Psi(r_1)\Psi(r_3)\cos(\Omega_3 - \Omega_1) + \Psi(r_2)\Psi(r_3)\cos(\Omega_3 - \Omega_2)) \end{aligned} \tag{2.22}$$

for the $N = 4$, or quadruple-slit interferometer the superposition probability becomes (Duarte 2014)

$$\begin{aligned} |\langle d|s\rangle|^2 &= \Psi(r_1)^2 + \Psi(r_2)^2 + \Psi(r_3)^2 + \Psi(r_4)^2 \\ &+ 2(\Psi(r_1)\Psi(r_2)\cos(\Omega_2 - \Omega_1) + \Psi(r_1)\Psi(r_3)\cos(\Omega_3 - \Omega_1) + \Psi(r_1)\Psi(r_4)\cos(\Omega_4 - \Omega_1) \\ &+ \Psi(r_2)\Psi(r_3)\cos(\Omega_3 - \Omega_2) + \Psi(r_2)\Psi(r_4)\cos(\Omega_4 - \Omega_2) + \Psi(r_3)\Psi(r_4)\cos(\Omega_4 - \Omega_3)) \end{aligned} \tag{2.23}$$

for the $N = 5$, or quintuple-slit interferometer the superposition probability becomes (Duarte 2014)

$$\begin{aligned} |\langle d|s\rangle|^2 &= \Psi(r_1)^2 + \Psi(r_2)^2 + \Psi(r_3)^2 + \Psi(r_4)^2 + \Psi(r_5)^2 \\ &+ 2(\Psi(r_1)\Psi(r_2)\cos(\Omega_2 - \Omega_1) + \Psi(r_1)\Psi(r_3)\cos(\Omega_3 - \Omega_1) + \Psi(r_1)\Psi(r_4)\cos(\Omega_4 - \Omega_1) \\ &+ \Psi(r_1)\Psi(r_5)\cos(\Omega_5 - \Omega_1) + \Psi(r_2)\Psi(r_3)\cos(\Omega_3 - \Omega_2) + \Psi(r_2)\Psi(r_4)\cos(\Omega_4 - \Omega_2) + \\ &+ \Psi(r_2)\Psi(r_5)\cos(\Omega_5 - \Omega_2) + \Psi(r_3)\Psi(r_4)\cos(\Omega_4 - \Omega_3) + \Psi(r_3)\Psi(r_5)\cos(\Omega_5 - \Omega_3) \\ &+ \Psi(r_4)\Psi(r_5)\cos(\Omega_5 - \Omega_4)) \end{aligned} \tag{2.24}$$

for the $N = 7$, or septuplet-slit interferometer the superposition probability becomes (Duarte 2014)

$$\begin{aligned} |\langle d|s\rangle|^2 &= \Psi(r_1)^2 + \Psi(r_2)^2 + \Psi(r_3)^2 + \Psi(r_4)^2 + \Psi(r_5)^2 + \Psi(r_6)^2 + \Psi(r_7)^2 \\ &+ 2(\Psi(r_1)\Psi(r_2)\cos(\Omega_2 - \Omega_1) + \Psi(r_1)\Psi(r_3)\cos(\Omega_3 - \Omega_1) + \Psi(r_1)\Psi(r_4)\cos(\Omega_4 - \Omega_1) \\ &+ \Psi(r_1)\Psi(r_5)\cos(\Omega_5 - \Omega_1) + \Psi(r_1)\Psi(r_6)\cos(\Omega_6 - \Omega_1) + \Psi(r_1)\Psi(r_7)\cos(\Omega_7 - \Omega_1) \\ &+ \Psi(r_2)\Psi(r_3)\cos(\Omega_3 - \Omega_2) + \Psi(r_2)\Psi(r_4)\cos(\Omega_4 - \Omega_2) + \Psi(r_2)\Psi(r_5)\cos(\Omega_5 - \Omega_2) \\ &+ \Psi(r_2)\Psi(r_6)\cos(\Omega_6 - \Omega_2) + \Psi(r_2)\Psi(r_7)\cos(\Omega_7 - \Omega_2) + \Psi(r_3)\Psi(r_4)\cos(\Omega_4 - \Omega_3) \\ &+ \Psi(r_3)\Psi(r_5)\cos(\Omega_5 - \Omega_3) + \Psi(r_3)\Psi(r_6)\cos(\Omega_6 - \Omega_3) + \Psi(r_3)\Psi(r_7)\cos(\Omega_7 - \Omega_3) \\ &+ \Psi(r_4)\Psi(r_5)\cos(\Omega_5 - \Omega_4) + \Psi(r_4)\Psi(r_6)\cos(\Omega_6 - \Omega_4) + \Psi(r_4)\Psi(r_7)\cos(\Omega_7 - \Omega_4) \\ &+ \Psi(r_5)\Psi(r_6)\cos(\Omega_6 - \Omega_5) + \Psi(r_5)\Psi(r_7)\cos(\Omega_7 - \Omega_5) + \Psi(r_6)\Psi(r_7)\cos(\Omega_7 - \Omega_6)) \end{aligned} \tag{2.25}$$

It should be re-emphasized that this explicit series of interferometric probabilities or the generalized superposition quantum probabilities given in equations (2.5), (2.12), and (2.14) can be applied to either explain measured interferometric profiles or to *predict* interferometric measurements well in advance. This has been demonstrated with populations of indistinguishable quanta (Duarte 1993, 1995). This also extends to illumination with a single photon where the pixels, or charges, will develop within the profiles allowed by these superposition quantum probabilities. The generalized interferometric probability equations have been programmed in Fortran 77, Visual Fortran, and Matlab (Duarte 1993, Duarte *et al* 2013). For calculations involving a large number of slits, up to $N = 2000$, Fortran is preferred. For measured and predicted interferometric profiles see chapter 10 and the cited references.

2.6 From quantum probabilities to measurable intensities

It should be reemphasized that the generalized interferometric probability equations, as given in equations (2.5), (2.14), (2.15), (2.16), and (2.17), *are probabilities.* As explained in the literature (Duarte 2004) these equations correspond to spatial probability distributions that determine the spatial boundaries for the arrival of quanta at the interference plane or detection surface *d*. The arrival of an individual photon, with an energy $E = h\nu$, is registered at the detector by the creation of a charge within the spatial boundaries of the probability distribution $\langle d|s\rangle\langle d|s\rangle^*$. For the case of populations of indistinguishable photons, a cumulative charge distribution within the spatial boundaries of $\langle d|s\rangle\langle d|s\rangle^*$ is registered at the detector (Duarte 2004). This cumulative charge distribution is what the experimenter observes as the spatial pattern of the photon count or the spatial distribution of the intensity.

Following Feynman *et al* (1965) and Duarte (2014, 2019) it can be shown that the *emission* intensity for a single photon is given by

$$\mathscr{I}(\nu) = \frac{\kappa\, h\nu}{\sigma\, T}|\gamma_{II\to I}|^2 \tag{2.26}$$

where κ is a numerical constant given by

$$\kappa = \frac{1}{4\pi^2} \tag{2.27}$$

while

$$\sigma = \frac{\mu^2}{4\pi\varepsilon_0 c\hbar} \tag{2.28}$$

is the emission cross section given in units of m^2, $h = 6.62607015 \times 10^{-34}$ Js is Planck's constant, ν is the emission frequency in Hz, T is the $II \to I$ transition time in s, and $|\gamma_{II\to I}|^2$ is the dimensionless transition probability.

For the particular case of an already emitted propagating photon the intensity, for single-photon propagation, equation (2.26) can be written as

$$\mathscr{I}(\nu) = K\, h\nu\, \langle d|s\rangle\langle d|s\rangle^* \tag{2.29}$$

where K is a constant characteristic of the emission, in units of $\mathrm{m^{-2}\ s^{-1}}$, and $\langle d|s\rangle\langle d|s\rangle^*$ is the interferometric spatial probability distribution (Duarte 2014).

At a fixed single frequency, equation (2.29) can also be expressed as

$$\mathscr{I}(\nu) \propto \langle d|s\rangle \langle d|s\rangle^* \tag{2.30}$$

For an ensemble of r indistinguishable quanta, equation (2.29) becomes

$$\mathscr{I}(\nu)_r = r\ K\ h\nu\ \langle d|s\rangle\langle d|s\rangle^* \tag{2.31}$$

It should be emphasized that in equations (2.29)–(2.31) it is the dimensionless probability $\langle d|s\rangle\langle d|s\rangle^*$ that provides all the spatial information. Also, the intensity $\mathscr{I}(\nu)$ has the units of $\mathrm{J\ s^{-1}\ m^{-2}}$ or $\mathrm{W\ m^{-2}}$. For further discussion on this topic see chapter 10 and appendix A.

2.7 Dirac's identities

Notation was very important in Dirac's physics. He wrote his equations in a meticulous and highly precise style. This is a brief introduction to Dirac's identities and the subject matter needs no previous introduction. It is simply an exercise in mathematical notation as invented, or discovered, by Dirac. These identities were originally presented in the various sections of Dirac's book and are compiled roughly in the same order as presented by Duarte (2014, 2019):

$$\langle\phi|\psi\rangle = \langle\psi|\phi\rangle^* \tag{2.32}$$

$$\langle\phi|\psi\rangle = \langle\phi|j\rangle\langle j|\psi\rangle \tag{2.33}$$

$$|\psi\rangle = |j\rangle\langle j|\psi\rangle \tag{2.34}$$

$$\langle\chi|A|\phi\rangle = \langle\chi|i\rangle\langle i|A|j\rangle\langle j|\phi\rangle \tag{2.35}$$

$$A = |i\rangle\langle i|A|j\rangle\langle j| \tag{2.36}$$

$$A|\phi\rangle = |i\rangle\langle i|A|j\rangle\langle j|\phi\rangle \tag{2.37}$$

If

$$\langle i|\phi\rangle = C_i \tag{2.38}$$

$$\langle i|\chi\rangle = D_i \tag{2.39}$$

then

$$|\phi\rangle = \sum_i |i\rangle C_i \tag{2.40}$$

$$|\chi\rangle = \sum_i |i\rangle D_i \tag{2.41}$$

and

$$\langle \chi | = \sum_i D_i^* \langle i | \tag{2.42}$$

Identity (2.42) is the abstracted expression of the probability amplitude

$$\langle \chi | \phi \rangle = \sum_{ij} D_j^* \langle j | i \rangle C_i \tag{2.43}$$

and using $\langle j | i \rangle = \delta_{ij}$

$$\langle \chi | \phi \rangle = \sum_i D_i^* C_i \tag{2.44}$$

As hinted by Dirac (Feynman *et al* 1965) identity (2.42) can be ultimately abstracted to

$$| = \sum_i | i \rangle \langle i | \tag{2.45}$$

The following Dirac identities are also important

$$|a\rangle|b\rangle = |b\rangle|a\rangle = |ab\rangle \tag{2.46}$$

$$|a\rangle|b\rangle|c\rangle\ldots = |abc\ldots\rangle \tag{2.47}$$

The commutative property is expressed as

$$|a\rangle|b\rangle = |b\rangle|a\rangle = |ab\rangle = |ba\rangle \tag{2.48}$$

Relevant to the quantum entanglement of several indistinguishable particles (Dirac 1958):

$$|a_1\rangle|b_2\rangle|c_3\rangle\ldots|g_n\rangle = |a_1b_2c_3\ldots g_n\rangle \tag{2.49}$$

$$|b_1\rangle|a_2\rangle|c_3\rangle\ldots|g_n\rangle = |b_1a_2c_3\ldots g_n\rangle \tag{2.50}$$

Identity (2.49) is equivalent to

$$|a\rangle_1|b\rangle_2|c\rangle_3\ldots|g\rangle_n = |a_1b_2c_3\ldots g_n\rangle \tag{2.51}$$

and introducing $|X\rangle$ to represent the ensemble $|a_1b_2c_3\ldots g_n\rangle$

$$|X\rangle = |a_1b_2c_3\ldots g_n\rangle \tag{2.52}$$

identity (2.52) is equivalent to

$$|X\rangle = |a\rangle_1|b\rangle_2|c\rangle_3\ldots|g\rangle_n \tag{2.53}$$

This identity is crucial to the derivation of the quantum entanglement probability amplitude from the interferometric foundations.

2.8 Quantum entanglement probability amplitudes for $n = N = 2$

The experimental diagram applicable to the fundamental configuration of quantum entanglement is depicted in figure 2.4. Here, s is a photon source of ν_1 and ν_2 emitted in the $+z$ and $-z$ directions, respectively. The quanta propagate along the optical axis, in the $+z$ and $-z$ directions, toward the polarization analyzers p_1 and p_2 toward the detectors d_1 and d_2, respectively.

The first essential, and experimentally sound, assumption is that the quanta emitted by the source are *indistinguishable*, that is $\nu_1 = \nu_2 = \nu$. Then, applying the Dirac interferometric principle stated in equation (2.2)

$$\langle d|s\rangle = \langle d|p_2\rangle\langle p_2|s\rangle + \langle d|p_1\rangle\langle p_1|s\rangle \tag{2.54}$$

Implicitly, this equation assumes that $d_1 = d_2 = d$ which is quite reasonable for a pair of matched detectors. Abstracting d from equation (2.54), this expression can be rewritten as

$$|s\rangle = |p_1\rangle\langle p_2|s\rangle + |p_1\rangle\langle p_1|s\rangle \tag{2.55}$$

Next, using the Dirac identity $|\psi\rangle = |j\rangle\langle j|\psi\rangle$ allows for the expression of

$$|D\rangle_2 = |p_2\rangle\langle p_2|s\rangle \tag{2.56}$$

and

$$|D\rangle_1 = |p_1\rangle\langle p_1|s\rangle \tag{2.57}$$

Substituting identities (2.56) and (2.57) into equation (2.55) yields

$$|s\rangle = (|D\rangle_2 + |D\rangle_1) \tag{2.58}$$

This equation is a probability amplitude $|s\rangle$, or state, representing the linear combination of two probability amplitudes, $|D\rangle_1$ and $|D\rangle_2$, and applies equally to an interferometric situation involving a single photon, or an ensemble of indistinguishable photons, and to the quantum entanglement framework being developed here. The crucial difference occurs at this very stage that invokes the use of the Dirac identity for 'similar quanta' as represented by

$$|X\rangle = |a\rangle_1|b\rangle_2|c\rangle_3\ldots|g\rangle_n$$

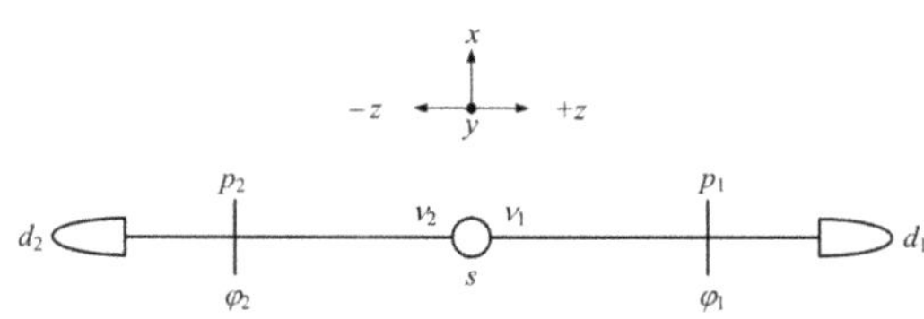

Figure 2.4. Experimental schematics for quantum entanglement. Here, s is photon source of ν_1 and ν_2 emitted in the $+z$- and $-z$-directions, respectively. The two linear polarization axes are indicated as x and y which are perpendicular to each other. The angles φ_1 and φ_2 are the polarization angles measured on a plane perpendicular to the propagation axis while d_1 and d_2 are the corresponding detectors (Duarte 2019).

This very step diverges from a purely interferometric situation given that the possibility of quanta 1 and quanta 2, is allowed with each quanta in different alternative states. That is, state $|x\rangle$ and state $|y\rangle$. Rewriting $|D\rangle_1$ and $|D\rangle_2$ in terms of polarization states as per identity (2.53), and using $|C\rangle_1$ and $|C\rangle_2$ to distinguish from a pure interferometric event, leads to

$$|s\rangle = (|C\rangle_2 + |C\rangle_1) \tag{2.59}$$

where

$$|C\rangle_2 = |x\rangle_1|y\rangle_2 \tag{2.60}$$

and

$$|C\rangle_1 = |y\rangle_1|x\rangle_2 \tag{2.61}$$

Inserting identities (2.60) and (2.61) into equation (2.59) leads directly to (Duarte 2013a, 2013b, 2014)

$$|s\rangle = (|x\rangle_1|y\rangle_2 + |y\rangle_1|x\rangle_2) \tag{2.62}$$

Following normalization, and designating $|s\rangle$ as $|\psi\rangle$, equation (2.62) can be expressed as

$$|\psi\rangle_+ = 2^{-1/2}(|x\rangle_1|y\rangle_2 + |y\rangle_1|x\rangle_2) \tag{2.63}$$

and its linear combination is

$$|\psi\rangle_- = 2^{-1/2}(|x\rangle_1|y\rangle_2 - |y\rangle_1|x\rangle_2) \tag{2.64}$$

which is the iconic probability amplitude for quantum entanglement first disclosed by Pryce and Ward (Pryce and Ward 1947, Ward 1949).

For notational functionality, equations (2.63) and (2.64) can be expressed, prior to normalization, in a 2 × 2 arrangement as (Duarte 2019)

$$\begin{matrix} +|C\rangle_1 + |C\rangle_2 \\ +|C\rangle_1 - |C\rangle_2 \end{matrix} \tag{2.65}$$

If the two quanta pairs have identical states of polarization, then equations (2.60) and (2.61) become

$$|C\rangle_2 = |x\rangle_1|x\rangle_2 \tag{2.66}$$

and

$$|C\rangle_1 = |y\rangle_1|y\rangle_2 \tag{2.67}$$

so that

$$|\psi\rangle^+ = 2^{-1/2}(|x\rangle_1|x\rangle_2 + |y\rangle_1|y\rangle_2) \tag{2.68}$$

and

$$|\psi\rangle^- = 2^{-1/2}(|x\rangle_1|x\rangle_2 - |y\rangle_1|y\rangle_2) \tag{2.69}$$

These types of probability amplitudes, $|\psi\rangle^+$ and $|\psi\rangle^-$, are of interest in quantum computing. In summary, the whole family of probability amplitudes relevant to two quanta is represented by $|\psi\rangle_+$, $|\psi\rangle_-$, $|\psi\rangle^+$, and $|\psi\rangle^-$.

2.9 Quantum entanglement probability amplitude for $n = N = 4$

The experimental diagram applicable to the $n = N = 4$ configuration for quantum entanglement is depicted in figure 2.5. Here, s is the photon source of two pairs of quanta emitted in opposite directions relative to each other: (ν_1, ν_2) and (ν_3, ν_4). In terms of conventional axis labeling (ν_1, ν_2) are emitted in the $+z$- and $-z$-directions whilst (ν_3, ν_4) are emitted in the $+x$- and $-x$-directions, respectively. In order to simplify the notation and axes labeling, in reference to figure 2.5, traditional axes labels are obviated and only polarization alternatives are assigned in pairs, in this case (x, y) along the $+z$- and $-z$-directions, and (φ, φ') along the corresponding orthogonal direction.

Again the first essential, and experimentally sound, assumption is that the quanta emitted by the source are *indistinguishable*, then assuming $d_1 = d_2 = d_3 = d_4 = d$ and applying the Dirac interferometric principle

$$\langle d|s\rangle = \langle d|p_4\rangle\langle p_4|s\rangle + \langle d|p_3\rangle\langle p_3|s\rangle + \langle d|p_2\rangle\langle p_2|s\rangle + \langle d|p_1\rangle\langle p_1|s\rangle \tag{2.70}$$

and following the procedure introduced in the previous section

$$|D\rangle_4 = |p_4\rangle\langle p_4|s\rangle \tag{2.71}$$

$$|D\rangle_3 = |p_3\rangle\langle p_3|s\rangle \tag{2.72}$$

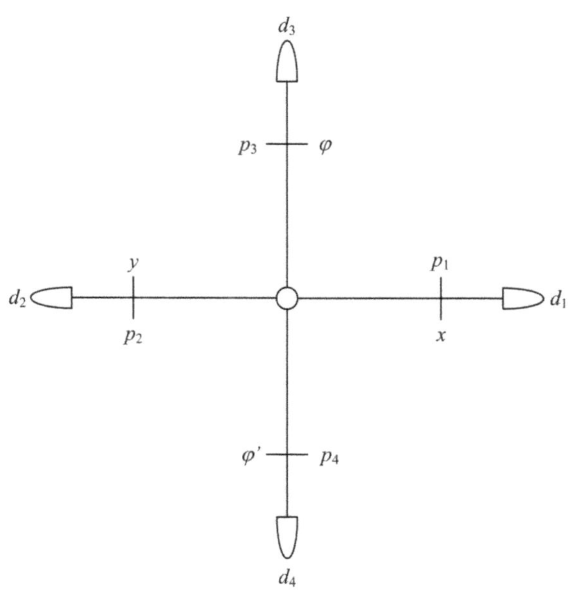

Figure 2.5. Schematics describing the experimental configuration to achieve quantum entanglement for $n = N = 2^2$ (Duarte 2019).

$$|D\rangle_2 = |p_2\rangle\langle p_2|s\rangle \tag{2.73}$$

$$|D\rangle_1 = |p_1\rangle\langle p_1|s\rangle \tag{2.74}$$

so that equation (2.70) becomes

$$|s\rangle = (|D\rangle_4 + |D\rangle_3 + |D\rangle_2 + |D\rangle_1) \tag{2.75}$$

Again, applying the Dirac identity (Dirac 1958) $|X\rangle = |a\rangle_1|b\rangle_2|c\rangle_3\ldots|g\rangle_n$ and using $|C\rangle_m$'s to differentiate from a pure interferometric situation

$$|C\rangle_1 = |x\rangle_1|y\rangle_2|\varphi\rangle_3|\varphi'\rangle_4 \tag{2.76}$$

$$|C\rangle_2 = |y\rangle_1|x\rangle_2|\varphi'\rangle_3|\varphi\rangle_4 \tag{2.77}$$

$$|C\rangle_3 = |\varphi\rangle_1|\varphi'\rangle_2|x\rangle_3|y\rangle_4 \tag{2.78}$$

$$|C\rangle_4 = |\varphi'\rangle_1|\varphi\rangle_2|y\rangle_3|x\rangle_4 \tag{2.79}$$

Next, introducing the normalized probability amplitude $|\psi\rangle_R$, where R is a *Roman numeral*, the relation between $|s\rangle$ and $|\psi\rangle_R$ is given by (Duarte 2015)

$$|\psi\rangle_R = N^{-1/2}|s\rangle \tag{2.80}$$

and the normalization equation for $N = 4$ is given by

$$1 = ||\psi\rangle_I|^2 + ||\psi\rangle_{II}|^2 + ||\psi\rangle_{III}|^2 + ||\psi\rangle_{IV}|^2 \tag{2.81}$$

The normalization condition leads to

$$|\psi\rangle_I = 4^{-1/2}(|C\rangle_1 + |C\rangle_2 + |C\rangle_3 + |C\rangle_4) \tag{2.82}$$

$$|\psi\rangle_{II} = 4^{-1/2}(|C\rangle_1 + |C\rangle_2 - |C\rangle_3 - |C\rangle_4) \tag{2.83}$$

$$|\psi\rangle_{III} = 4^{-1/2}(|C\rangle_1 - |C\rangle_2 + |C\rangle_3 - |C\rangle_4) \tag{2.84}$$

$$|\psi\rangle_{IV} = 4^{-1/2}(|C\rangle_1 - |C\rangle_2 - |C\rangle_3 + |C\rangle_4) \tag{2.85}$$

Thus, using identities (2.76)–(2.79) into equations (2.82)–(2.85), the explicit expressions for the probability amplitudes become

$$\begin{aligned}|\psi\rangle_I = 4^{-1/2}(&|x\rangle_1|y\rangle_2|\varphi\rangle_3|\varphi'\rangle_4 + |y\rangle_1|x\rangle_2|\varphi'\rangle_3|\varphi\rangle_4 \\ &+ |\varphi\rangle_1|\varphi'\rangle_2|x\rangle_3|y\rangle_4 + |\varphi'\rangle_1|\varphi\rangle_2|y\rangle_3|x\rangle_4)\end{aligned} \tag{2.86}$$

$$\begin{aligned}|\psi\rangle_{II} = 4^{-1/2}(&|x\rangle_1|y\rangle_2|\varphi\rangle_3|\varphi'\rangle_4 + |y\rangle_1|x\rangle_2|\varphi'\rangle_3|\varphi\rangle_4 \\ &- |\varphi\rangle_1|\varphi'\rangle_2|x\rangle_3|y\rangle_4 + |\varphi'\rangle_1|\varphi\rangle_2|y\rangle_3|x\rangle_4)\end{aligned} \tag{2.87}$$

$$\begin{aligned}|\psi\rangle_{III} = 4^{-1/2}(&|x\rangle_1|y\rangle_2|\varphi\rangle_3|\varphi'\rangle_4 - |y\rangle_1|x\rangle_2|\varphi'\rangle_3|\varphi\rangle_4 \\ &+ |\varphi\rangle_1|\varphi'\rangle_2|x\rangle_3|y\rangle_4 - |\varphi'\rangle_1|\varphi\rangle_2|y\rangle_3|x\rangle_4)\end{aligned} \tag{2.88}$$

$$|\psi\rangle_{IV} = 4^{-1/2}(|x\rangle_1|y\rangle_2|\varphi\rangle_3|\varphi'\rangle_4 - |y\rangle_1|x\rangle_2|\varphi'\rangle_3|\varphi\rangle_4 \\ - |\varphi\rangle_1|\varphi'\rangle_2|x\rangle_3|y\rangle_4 + |\varphi'\rangle_1|\varphi\rangle_2|y\rangle_3|x\rangle_4) \tag{2.89}$$

It should be noted that the symmetry pattern in the sign convention of the $|C\rangle_m$ states can be best appreciated in the 4 × 4 arrangement (Duarte 2018, 2019)

$$\begin{aligned} &+|C\rangle_1 + |C\rangle_2 + |C\rangle_3 + |C\rangle_4 \\ &+|C\rangle_1 + |C\rangle_2 - |C\rangle_3 - |C\rangle_4 \\ &+|C\rangle_1 - |C\rangle_2 + |C\rangle_3 - |C\rangle_4 \\ &+|C\rangle_1 - |C\rangle_2 - |C\rangle_3 + |C\rangle_4 \end{aligned} \tag{2.90}$$

2.10 Quantum entanglement probability amplitudes for $n = N = 8$

The experimental diagram applicable to the $n = N = 8$ configuration of quantum entanglement is depicted in figure 2.6. Four axes are involved, each adjacent axis at $\theta = \pi/4$ rad relative to each other. Since each axis has a + and a − direction, then $N = 8$ channels of propagation are available. Here, s is the photon source of *four pairs* of quanta emitted in opposite directions relative to each other. Again, in order to simplify the notation and axes labeling, in reference to figure 2.6, axes labels are obviated and only polarization alternatives are assigned in pairs, in this case (x, y) along the $+z$- and $-z$-directions, and (φ, φ') along the corresponding orthogonal direction. The first axis at $\theta = \pi/4$, relative to the z-axis, involves the (ϕ, ϕ') polarizations and the axis orthogonal to it involves the (ϑ, ϑ') polarizations, as illustrated in figure 2.6.

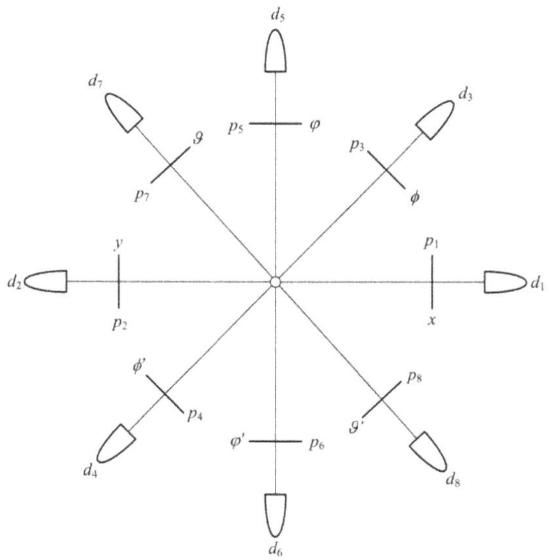

Figure 2.6. Schematics describing the experimental configuration to achieve quantum entanglement for $n = N = 2^3$ (Duarte 2019).

Following an identical methodology as in the previous sections (Duarte 2019)

$$|s\rangle = |D\rangle_8 + |D\rangle_7 + |D\rangle_6 + |D\rangle_5 + |D\rangle_4 + |D\rangle_3 + |D\rangle_2 + |D\rangle_1 \tag{2.91}$$

Again, applying the Dirac identity (Dirac 1958) $|X\rangle = |a\rangle_1|b\rangle_2|c\rangle_3\ldots|g\rangle_n$ and using $|C\rangle_m$'s to differentiate from a pure interferometric situation

$$|C\rangle_1 = |x\rangle_1|y\rangle_2|\varphi\rangle_3|\varphi'\rangle_4|\phi\rangle_5|\phi'\rangle_6|\vartheta\rangle_7|\vartheta'\rangle_8 \tag{2.92}$$

$$|C\rangle_2 = |y\rangle_1|x\rangle_2|\varphi'\rangle_3|\varphi\rangle_4|\phi'\rangle_5|\phi\rangle_6|\vartheta'\rangle_7|\vartheta\rangle_8 \tag{2.93}$$

$$|C\rangle_3 = |\varphi\rangle_1|\varphi'\rangle_2|\phi\rangle_3|\phi'\rangle_4|\vartheta\rangle_5|\vartheta'\rangle_6|x\rangle_7|y\rangle_8 \tag{2.94}$$

$$|C\rangle_4 = |\varphi'\rangle_1|\varphi\rangle_2|\phi'\rangle_3|\phi\rangle_4|\vartheta'\rangle_5|\vartheta\rangle_6|y\rangle_7|x\rangle_8 \tag{2.95}$$

$$|C\rangle_5 = |\phi\rangle_1|\phi'\rangle_2|\vartheta\rangle_3|\vartheta'\rangle_4|x\rangle_5|y\rangle_6|\varphi\rangle_7|\varphi'\rangle_8 \tag{2.96}$$

$$|C\rangle_6 = |\phi'\rangle_1|\phi\rangle_2|\vartheta'\rangle_3|\vartheta\rangle_4|y\rangle_5|x\rangle_6|\varphi'\rangle_7|\varphi\rangle_8 \tag{2.97}$$

$$|C\rangle_7 = |\vartheta\rangle_1|\vartheta'\rangle_2|x\rangle_3|y\rangle_4|\varphi\rangle_5|\varphi'\rangle_6|\phi\rangle_7|\phi'\rangle_8 \tag{2.98}$$

$$|C\rangle_8 = |\vartheta'\rangle_1|\vartheta\rangle_2|y\rangle_3|x\rangle_4|\varphi'\rangle_5|\varphi\rangle_6|\phi'\rangle_7|\phi\rangle_8 \tag{2.99}$$

Equations (2.92)–(2.99) describe quanta pairs (1, 2), (3, 4), (5, 6), (7, 8) in a particular polarization order as outlined by $|C\rangle_1$. Next, within pairs, polarizations are interchanged as depicted in $|C\rangle_2$. In $|C\rangle_3$, the second pair moves to the position of the first pair which is displaced to the position of the last pair. Transposition of the polarizations follows as indicated in $|C\rangle_4$, and so on (Duarte 2018). The only constant is the position of the quanta 1, 2, 3, 4, 5, 6, 7, 8.

For $n = N = 2^3$ the normalized probability amplitudes become

$$|\psi\rangle_I = 8^{-1/2}(|C\rangle_1 + |C\rangle_2 + |C\rangle_3 + |C\rangle_4 + |C\rangle_5 + |C\rangle_6 + |C\rangle_7 + |C\rangle_8) \tag{2.100}$$

$$|\psi\rangle_{II} = 8^{-1/2}(|C\rangle_1 + |C\rangle_2 + |C\rangle_3 + |C\rangle_4 - |C\rangle_5 - |C\rangle_6 - |C\rangle_7 - |C\rangle_8) \tag{2.101}$$

$$|\psi\rangle_{III} = 8^{-1/2}(|C\rangle_1 + |C\rangle_2 - |C\rangle_3 - |C\rangle_4 + |C\rangle_5 + |C\rangle_6 - |C\rangle_7 - |C\rangle_8) \tag{2.102}$$

$$|\psi\rangle_{IV} = 8^{-1/2}(|C\rangle_1 + |C\rangle_2 - |C\rangle_3 - |C\rangle_4 - |C\rangle_5 - |C\rangle_6 + |C\rangle_7 + |C\rangle_8) \tag{2.103}$$

$$|\psi\rangle_V = 8^{-1/2}(|C\rangle_1 - |C\rangle_2 + |C\rangle_3 - |C\rangle_4 + |C\rangle_5 - |C\rangle_6 + |C\rangle_7 - |C\rangle_8) \tag{2.104}$$

$$|\psi\rangle_{VI} = 8^{-1/2}(|C\rangle_1 - |C\rangle_2 + |C\rangle_3 - |C\rangle_4 - |C\rangle_5 + |C\rangle_6 - |C\rangle_7 + |C\rangle_8) \tag{2.105}$$

$$|\psi\rangle_{VII} = 8^{-1/2}(|C\rangle_1 - |C\rangle_2 - |C\rangle_3 + |C\rangle_4 + |C\rangle_5 - |C\rangle_6 - |C\rangle_7 + |C\rangle_8) \tag{2.106}$$

$$|\psi\rangle_{VIII} = 8^{-1/2}(|C\rangle_1 - |C\rangle_2 - |C\rangle_3 + |C\rangle_4 - |C\rangle_5 + |C\rangle_6 + |C\rangle_7 - |C\rangle_8) \tag{2.107}$$

The sign alternatives in this system of equations can be better appreciated if the $|C\rangle_m$ amplitudes are expressed in an 8 × 8 format with the *bra ket* (|⟩) symbols abstracted (Duarte 2018, 2019)

$$\begin{aligned}
&+C_1+C_2+C_3+C_4+C_5+C_6+C_7+C_8\\
&+C_1+C_2+C_3+C_4-C_5-C_6-C_7-C_8\\
&+C_1+C_2-C_3-C_4+C_5+C_6-C_7-C_8\\
&+C_1+C_2-C_3-C_4-C_5-C_6+C_7+C_8\\
&+C_1-C_2+C_3-C_4+C_5-C_6+C_7-C_8\\
&+C_1-C_2+C_3-C_4-C_5+C_6-C_7+C_8\\
&+C_1-C_2-C_3+C_4+C_5-C_6-C_7+C_8\\
&+C_1-C_2-C_3+C_4-C_5+C_6+C_7-C_8
\end{aligned} \tag{2.108}$$

2.11 Quantum entanglement probability amplitudes for $n = N = 16$

For $n = N = 2^4$

$$|\psi\rangle_I = 16^{-1/2}(|C\rangle_1 + |C\rangle_2 + |C\rangle_3 + |C\rangle_4 + |C\rangle_5 + |C\rangle_6 + |C\rangle_7 + |C\rangle_8 + |C\rangle_9 + |C\rangle_{10} + |C\rangle_{11} + |C\rangle_{12} + |C\rangle_{13} + |C\rangle_{14} + |C\rangle_{15} + |C\rangle_{16}) \tag{2.109}$$

and so on for $|\psi\rangle_{II}, |\psi\rangle_{III}, |\psi\rangle_{IV} \ldots |\psi\rangle_{XVI}$. The sign alternatives for the $|C\rangle_m$ probability amplitudes, with the *bra ket* (|⟩) symbols abstracted, are highlighted in the following 16 × 16 array (Duarte 2018, 2019)

$$\begin{aligned}
&+C_1+C_2+C_3+C_4+C_5+C_6+C_7+C_8+C_9+C_{10}+C_{11}+C_{12}+C_{13}+C_{14}+C_{15}+C_{16}\\
&+C_1+C_2+C_3+C_4+C_5+C_6+C_7+C_8-C_9-C_{10}-C_{11}-C_{12}-C_{13}-C_{14}-C_{15}-C_{16}\\
&+C_1+C_2+C_3+C_4-C_5-C_6-C_7-C_8+C_9+C_{10}+C_{11}+C_{12}-C_{13}-C_{14}-C_{15}-C_{16}\\
&+C_1+C_2+C_3+C_4-C_5-C_6-C_7-C_8-C_9-C_{10}-C_{11}-C_{12}+C_{13}+C_{14}+C_{15}+C_{16}\\
&+C_1+C_2-C_3-C_4+C_5+C_6-C_7-C_8+C_9+C_{10}-C_{11}-C_{12}+C_{13}+C_{14}-C_{15}-C_{16}\\
&+C_1+C_2-C_3-C_4+C_5+C_6-C_7-C_8-C_9-C_{10}+C_{11}+C_{12}-C_{13}-C_{14}+C_{15}+C_{16}\\
&+C_1+C_2-C_3-C_4-C_5-C_6+C_7+C_8+C_9+C_{10}-C_{11}-C_{12}-C_{13}-C_{14}+C_{15}+C_{16}\\
&+C_1+C_2-C_3-C_4-C_5-C_6+C_7+C_8-C_9-C_{10}+C_{11}+C_{12}+C_{13}+C_{14}-C_{15}-C_{16}\\
&+C_1-C_2+C_3-C_4+C_5-C_6+C_7-C_8+C_9-C_{10}+C_{11}-C_{12}+C_{13}-C_{14}+C_{15}-C_{16}\\
&+C_1-C_2+C_3-C_4+C_5-C_6+C_7-C_8-C_9+C_{10}-C_{11}+C_{12}-C_{13}+C_{14}-C_{15}+C_{16}\\
&+C_1-C_2+C_3-C_4-C_5+C_6-C_7+C_8+C_9-C_{10}+C_{11}-C_{12}-C_{13}+C_{14}-C_{15}+C_{16}\\
&+C_1-C_2+C_3-C_4-C_5+C_6-C_7+C_8-C_9+C_{10}-C_{11}+C_{12}+C_{13}-C_{14}+C_{15}-C_{16}\\
&+C_1-C_2-C_3+C_4+C_5-C_6-C_7+C_8+C_9-C_{10}-C_{11}+C_{12}+C_{13}-C_{14}-C_{15}+C_{16}\\
&+C_1-C_2-C_3+C_4+C_5-C_6-C_7+C_8-C_9+C_{10}+C_{11}-C_{12}-C_{13}+C_{14}+C_{15}-C_{16}\\
&+C_1-C_2-C_3+C_4-C_5+C_6+C_7-C_8+C_9-C_{10}-C_{11}+C_{12}-C_{13}+C_{14}+C_{15}-C_{16}\\
&+C_1-C_2-C_3+C_4-C_5+C_6+C_7-C_8-C_9+C_{10}+C_{11}-C_{12}+C_{13}-C_{14}-C_{15}+C_{16}
\end{aligned} \tag{2.110}$$

2.12 Quantum entanglement probability amplitudes for $n = N = 2^1, 2^2, 2^3, 2^4, \ldots, 2^r$

First, the probability amplitude $|s\rangle$ can generally be expressed as

$$|s\rangle = \sum_{j=1}^{N} (\pm)|C\rangle_{N+1-j} \tag{2.111}$$

and for notational consistency the normalized probability amplitude is expressed as

$$|\psi\rangle_R = N^{-1/2}|s\rangle \tag{2.112}$$

which means

$$|\psi\rangle_R = N^{-1/2}\sum_{j=1}^{N}(\pm)|C\rangle_{N+1-j} \tag{2.113}$$

where, the subscript R is a *Roman numeral*, the $N^{-1/2}$ factor is derived from the normalization operation

$$1 = ||\psi\rangle_I|^2 + ||\psi\rangle_{II}|^2 + ||\psi\rangle_{III}|^2 + ||\psi\rangle_{IV}|^2 + \ldots \tag{2.114}$$

and the sign alternative $(\pm)$ means that every possible sign alternative in the series must be considered.

Next, using the Dirac's identity for similar quanta, the individual path probability amplitudes $|C\rangle_{N+1-j}$ are expressed in a generalized form as (Duarte 2016, 2018)

$$|C\rangle_{N+1-j} = \prod_{m=1,3,5\ldots}^{n} |a\rangle_m|b\rangle_{m+1} \tag{2.115}$$

Equation (2.113) is a mathematical expression for a series of superposition probability amplitudes beginning at $|C\rangle_{N+1-j}$ with $j = 1$, and ending at $|C\rangle_1$, which is reached when $j = N$. In this notation, n is the total number of quanta, which is an even number since quanta participate in pairs. For each pair, (1, 2), (3, 4)…$(m, m + 1)$, $|a\rangle_m$, $|b\rangle_{m+1}$ represent a set of orthogonal polarization alternatives such as (x, y), (φ, φ'), and so on (Duarte 2018).

Explicitly, for $n = N = 2^3$, and for the values of j expressed in parenthesis, equation (2.113) yields:

$(j = N - 0)$:

$$|C\rangle_1 = |x\rangle_1|y\rangle_2|\varphi\rangle_3|\varphi'\rangle_4|\phi\rangle_5|\phi'\rangle_6|\vartheta\rangle_7|\vartheta'\rangle_8 \tag{2.116}$$

$(j = N - 1)$:

$$|C\rangle_2 = |y\rangle_1|x\rangle_2|\varphi'\rangle_3|\varphi\rangle_4|\phi'\rangle_5|\phi\rangle_6|\vartheta'\rangle_7|\vartheta\rangle_8 \tag{2.117}$$

$(j = N - 2)$:

$$|C\rangle_3 = |\varphi\rangle_1|\varphi'\rangle_2|\phi\rangle_3|\phi'\rangle_4|\vartheta\rangle_5|\vartheta'\rangle_6|x\rangle_7|y\rangle_8 \tag{2.118}$$

$(j = N - 3)$:

$$|C\rangle_4 = |\varphi'\rangle_1|\varphi\rangle_2|\phi'\rangle_3|\phi\rangle_4|\vartheta'\rangle_5|\vartheta\rangle_6|y\rangle_7|x\rangle_8 \tag{2.119}$$

$(j = N - 4)$:

$$|C\rangle_5 = |\phi\rangle_1|\phi'\rangle_2|\vartheta\rangle_3|\vartheta'\rangle_4|x\rangle_5|y\rangle_6|\varphi\rangle_7|\varphi'\rangle_8 \tag{2.120}$$

$(j = N - 5)$:

$$|C\rangle_6 = |\phi'\rangle_1|\phi\rangle_2|\vartheta'\rangle_3|\vartheta\rangle_4|y\rangle_5|x\rangle_6|\varphi'\rangle_7|\varphi\rangle_8 \tag{2.121}$$

$(j = N - 6)$:

$$|C\rangle_7 = |\vartheta\rangle_1|\vartheta'\rangle_2|x\rangle_3|y\rangle_4|\varphi\rangle_5|\varphi'\rangle_6|\phi\rangle_7|\phi'\rangle_8 \tag{2.122}$$

$(j = N - 7)$:

$$|C\rangle_8 = |\vartheta'\rangle_1|\vartheta\rangle_2|y\rangle_3|x\rangle_4|\varphi'\rangle_5|\varphi\rangle_6|\phi'\rangle_7|\phi\rangle_8 \tag{2.123}$$

Substitution of the corresponding $|C\rangle_{N+1-j}$ amplitudes, that is $|C\rangle_1$, $|C\rangle_2$, $|C\rangle_3 \ldots |C\rangle_8$ (from equations (2.116)–(2.123)), into equations (2.100)–(2.107), which are also directly obtainable from equation (2.111), leads to the explicit series of probability amplitudes $|\psi\rangle_I$, $|\psi\rangle_{II}$, $|\psi\rangle_{III} \ldots |\psi\rangle_{VIII}$ in terms of the polarization coordinates (x, y), (φ, φ'), (ϕ, ϕ'), and (ϑ, ϑ') (Duarte and Taylor 2017)

$$\begin{aligned} |\psi\rangle_I = 8^{-1/2}(&|x\rangle_1|y\rangle_2|\varphi\rangle_3|\varphi'\rangle_4|\phi\rangle_5|\phi'\rangle_6|\vartheta\rangle_7|\vartheta'\rangle_8 \\ &+ |y\rangle_1|x\rangle_2|\varphi'\rangle_3|\varphi\rangle_4|\phi'\rangle_5|\phi\rangle_6|\vartheta'\rangle_7|\vartheta\rangle_8 \\ &+ |\varphi\rangle_1|\varphi'\rangle_2|\phi\rangle_3|\phi'\rangle_4|\vartheta\rangle_5|\vartheta'\rangle_6|x\rangle_7|y\rangle_8 \\ &+ |\varphi'\rangle_1|\varphi\rangle_2|\phi'\rangle_3|\phi\rangle_4|\vartheta'\rangle_5|\vartheta\rangle_6|y\rangle_7|x\rangle_8 \\ &+ |\phi\rangle_1|\phi'\rangle_2|\vartheta\rangle_3|\vartheta'\rangle_4|x\rangle_5|y\rangle_6|\varphi\rangle_7|\varphi'\rangle_8 \\ &+ |\phi'\rangle_1|\phi\rangle_2|\vartheta'\rangle_3|\vartheta\rangle_4|y\rangle_5|x\rangle_6|\varphi'\rangle_7|\varphi\rangle_8 \\ &+ |\vartheta\rangle_1|\vartheta'\rangle_2|x\rangle_3|y\rangle_4|\varphi\rangle_5|\varphi'\rangle_6|\phi\rangle_7|\phi'\rangle_8 \\ &+ |\vartheta'\rangle_1|\vartheta\rangle_2|y\rangle_3|x\rangle_4|\varphi'\rangle_5|\varphi\rangle_6|\phi'\rangle_7|\phi\rangle_8) \end{aligned} \tag{2.124}$$

$$\begin{aligned} |\psi\rangle_{II} = 8^{-1/2}(&|x\rangle_1|y\rangle_2|\varphi\rangle_3|\varphi'\rangle_4|\phi\rangle_5|\phi'\rangle_6|\vartheta\rangle_7|\vartheta'\rangle_8 \\ &+ |y\rangle_1|x\rangle_2|\varphi'\rangle_3|\varphi\rangle_4|\phi'\rangle_5|\phi\rangle_6|\vartheta'\rangle_7|\vartheta\rangle_8 \\ &+ |\varphi\rangle_1|\varphi'\rangle_2|\phi\rangle_3|\phi'\rangle_4|\vartheta\rangle_5|\vartheta'\rangle_6|x\rangle_7|y\rangle_8 \\ &+ |\varphi'\rangle_1|\varphi\rangle_2|\phi'\rangle_3|\phi\rangle_4|\vartheta'\rangle_5|\vartheta\rangle_6|y\rangle_7|x\rangle_8 \\ &- |\phi\rangle_1|\phi'\rangle_2|\vartheta\rangle_3|\vartheta'\rangle_4|x\rangle_5|y\rangle_6|\varphi\rangle_7|\varphi'\rangle_8 \\ &- |\phi'\rangle_1|\phi\rangle_2|\vartheta'\rangle_3|\vartheta\rangle_4|y\rangle_5|x\rangle_6|\varphi'\rangle_7|\varphi\rangle_8 \\ &- |\vartheta\rangle_1|\vartheta'\rangle_2|x\rangle_3|y\rangle_4|\varphi\rangle_5|\varphi'\rangle_6|\phi\rangle_7|\phi'\rangle_8 \\ &- |\vartheta'\rangle_1|\vartheta\rangle_2|y\rangle_3|x\rangle_4|\varphi'\rangle_5|\varphi\rangle_6|\phi'\rangle_7|\phi\rangle_8) \end{aligned} \tag{2.125}$$

$$\begin{aligned} |\psi\rangle_{III} = 8^{-1/2}(&|x\rangle_1|y\rangle_2|\varphi\rangle_3|\varphi'\rangle_4|\phi\rangle_5|\phi'\rangle_6|\vartheta\rangle_7|\vartheta'\rangle_8 \\ &+ |y\rangle_1|x\rangle_2|\varphi'\rangle_3|\varphi\rangle_4|\phi'\rangle_5|\phi\rangle_6|\vartheta'\rangle_7|\vartheta\rangle_8 \\ &- |\varphi\rangle_1|\varphi'\rangle_2|\phi\rangle_3|\phi'\rangle_4|\vartheta\rangle_5|\vartheta'\rangle_6|x\rangle_7|y\rangle_8 \\ &- |\varphi'\rangle_1|\varphi\rangle_2|\phi'\rangle_3|\phi\rangle_4|\vartheta'\rangle_5|\vartheta\rangle_6|y\rangle_7|x\rangle_8 \\ &+ |\phi\rangle_1|\phi'\rangle_2|\vartheta\rangle_3|\vartheta'\rangle_4|x\rangle_5|y\rangle_6|\varphi\rangle_7|\varphi'\rangle_8 \\ &+ |\phi'\rangle_1|\phi\rangle_2|\vartheta'\rangle_3|\vartheta\rangle_4|y\rangle_5|x\rangle_6|\varphi'\rangle_7|\varphi\rangle_8 \\ &- |\vartheta\rangle_1|\vartheta'\rangle_2|x\rangle_3|y\rangle_4|\varphi\rangle_5|\varphi'\rangle_6|\phi\rangle_7|\phi'\rangle_8 \\ &- |\vartheta'\rangle_1|\vartheta\rangle_2|y\rangle_3|x\rangle_4|\varphi'\rangle_5|\varphi\rangle_6|\phi'\rangle_7|\phi\rangle_8) \end{aligned} \tag{2.126}$$

$$\begin{aligned}|\psi\rangle_{IV} = 8^{-1/2}(&|x\rangle_1|y\rangle_2|\varphi\rangle_3|\varphi'\rangle_4|\phi\rangle_5|\phi'\rangle_6|\vartheta\rangle_7|\vartheta'\rangle_8\\ &+|y\rangle_1|x\rangle_2|\varphi'\rangle_3|\varphi\rangle_4|\phi'\rangle_5|\phi\rangle_6|\vartheta'\rangle_7|\vartheta\rangle_8\\ &-|\varphi\rangle_1|\varphi'\rangle_2|\phi\rangle_3|\phi'\rangle_4|\vartheta\rangle_5|\vartheta'\rangle_6|x\rangle_7|y\rangle_8\\ &-|\varphi'\rangle_1|\varphi\rangle_2|\phi'\rangle_3|\phi\rangle_4|\vartheta'\rangle_5|\vartheta\rangle_6|y\rangle_7|x\rangle_8\\ &-|\phi\rangle_1|\phi'\rangle_2|\vartheta\rangle_3|\vartheta'\rangle_4|x\rangle_5|y\rangle_6|\varphi\rangle_7|\varphi'\rangle_8\\ &-|\phi'\rangle_1|\phi\rangle_2|\vartheta'\rangle_3|\vartheta\rangle_4|y\rangle_5|x\rangle_6|\varphi'\rangle_7|\varphi\rangle_8\\ &+|\vartheta\rangle_1|\vartheta'\rangle_2|x\rangle_3|y\rangle_4|\varphi\rangle_5|\varphi'\rangle_6|\phi\rangle_7|\phi'\rangle_8\\ &+|\vartheta'\rangle_1|\vartheta\rangle_2|y\rangle_3|x\rangle_4|\varphi'\rangle_5|\varphi\rangle_6|\phi'\rangle_7|\phi\rangle_8)\end{aligned} \tag{2.127}$$

$$\begin{aligned}|\psi\rangle_{V} = 8^{-1/2}(&|x\rangle_1|y\rangle_2|\varphi\rangle_3|\varphi'\rangle_4|\phi\rangle_5|\phi'\rangle_6|\vartheta\rangle_7|\vartheta'\rangle_8\\ &-|y\rangle_1|x\rangle_2|\varphi'\rangle_3|\varphi\rangle_4|\phi'\rangle_5|\phi\rangle_6|\vartheta'\rangle_7|\vartheta\rangle_8\\ &+|\varphi\rangle_1|\varphi'\rangle_2|\phi\rangle_3|\phi'\rangle_4|\vartheta\rangle_5|\vartheta'\rangle_6|x\rangle_7|y\rangle_8\\ &-|\varphi'\rangle_1|\varphi\rangle_2|\phi'\rangle_3|\phi\rangle_4|\vartheta'\rangle_5|\vartheta\rangle_6|y\rangle_7|x\rangle_8\\ &+|\phi\rangle_1|\phi'\rangle_2|\vartheta\rangle_3|\vartheta'\rangle_4|x\rangle_5|y\rangle_6|\varphi\rangle_7|\varphi'\rangle_8\\ &-|\phi'\rangle_1|\phi\rangle_2|\vartheta'\rangle_3|\vartheta\rangle_4|y\rangle_5|x\rangle_6|\varphi'\rangle_7|\varphi\rangle_8\\ &+|\vartheta\rangle_1|\vartheta'\rangle_2|x\rangle_3|y\rangle_4|\varphi\rangle_5|\varphi'\rangle_6|\phi\rangle_7|\phi'\rangle_8\\ &-|\vartheta'\rangle_1|\vartheta\rangle_2|y\rangle_3|x\rangle_4|\varphi'\rangle_5|\varphi\rangle_6|\phi'\rangle_7|\phi\rangle_8)\end{aligned} \tag{2.128}$$

$$\begin{aligned}|\psi\rangle_{VI} = 8^{-1/2}(&|x\rangle_1|y\rangle_2|\varphi\rangle_3|\varphi'\rangle_4|\phi\rangle_5|\phi'\rangle_6|\vartheta\rangle_7|\vartheta'\rangle_8\\ &-|y\rangle_1|x\rangle_2|\varphi'\rangle_3|\varphi\rangle_4|\phi'\rangle_5|\phi\rangle_6|\vartheta'\rangle_7|\vartheta\rangle_8\\ &+|\varphi\rangle_1|\varphi'\rangle_2|\phi\rangle_3|\phi'\rangle_4|\vartheta\rangle_5|\vartheta'\rangle_6|x\rangle_7|y\rangle_8\\ &-|\varphi'\rangle_1|\varphi\rangle_2|\phi'\rangle_3|\phi\rangle_4|\vartheta'\rangle_5|\vartheta\rangle_6|y\rangle_7|x\rangle_8\\ &-|\phi\rangle_1|\phi'\rangle_2|\vartheta\rangle_3|\vartheta'\rangle_4|x\rangle_5|y\rangle_6|\varphi\rangle_7|\varphi'\rangle_8\\ &+|\phi'\rangle_1|\phi\rangle_2|\vartheta'\rangle_3|\vartheta\rangle_4|y\rangle_5|x\rangle_6|\varphi'\rangle_7|\varphi\rangle_8\\ &-|\vartheta\rangle_1|\vartheta'\rangle_2|x\rangle_3|y\rangle_4|\varphi\rangle_5|\varphi'\rangle_6|\phi\rangle_7|\phi'\rangle_8\\ &+|\vartheta'\rangle_1|\vartheta\rangle_2|y\rangle_3|x\rangle_4|\varphi'\rangle_5|\varphi\rangle_6|\phi'\rangle_7|\phi\rangle_8)\end{aligned} \tag{2.129}$$

$$\begin{aligned}|\psi\rangle_{VII} = 8^{-1/2}(&|x\rangle_1|y\rangle_2|\varphi\rangle_3|\varphi'\rangle_4|\phi\rangle_5|\phi'\rangle_6|\vartheta\rangle_7|\vartheta'\rangle_8\\ &-|y\rangle_1|x\rangle_2|\varphi'\rangle_3|\varphi\rangle_4|\phi'\rangle_5|\phi\rangle_6|\vartheta'\rangle_7|\vartheta\rangle_8\\ &-|\varphi\rangle_1|\varphi'\rangle_2|\phi\rangle_3|\phi'\rangle_4|\vartheta\rangle_5|\vartheta'\rangle_6|x\rangle_7|y\rangle_8\\ &+|\varphi'\rangle_1|\varphi\rangle_2|\phi'\rangle_3|\phi\rangle_4|\vartheta'\rangle_5|\vartheta\rangle_6|y\rangle_7|x\rangle_8\\ &+|\phi\rangle_1|\phi'\rangle_2|\vartheta\rangle_3|\vartheta'\rangle_4|x\rangle_5|y\rangle_6|\varphi\rangle_7|\varphi'\rangle_8\\ &-|\phi'\rangle_1|\phi\rangle_2|\vartheta'\rangle_3|\vartheta\rangle_4|y\rangle_5|x\rangle_6|\varphi'\rangle_7|\varphi\rangle_8\\ &-|\vartheta\rangle_1|\vartheta'\rangle_2|x\rangle_3|y\rangle_4|\varphi\rangle_5|\varphi'\rangle_6|\phi\rangle_7|\phi'\rangle_8\\ &+|\vartheta'\rangle_1|\vartheta\rangle_2|y\rangle_3|x\rangle_4|\varphi'\rangle_5|\varphi\rangle_6|\phi'\rangle_7|\phi\rangle_8)\end{aligned} \tag{2.130}$$

$$\begin{aligned}|\psi\rangle_{VIII} = 8^{-1/2}(&|x\rangle_1|y\rangle_2|\varphi\rangle_3|\varphi'\rangle_4|\phi\rangle_5|\phi'\rangle_6|\vartheta\rangle_7|\vartheta'\rangle_8 \\ &- |y\rangle_1|x\rangle_2|\varphi'\rangle_3|\varphi\rangle_4|\phi'\rangle_5|\phi\rangle_6|\vartheta'\rangle_7|\vartheta\rangle_8 \\ &- |\varphi\rangle_1|\varphi'\rangle_2|\phi\rangle_3|\phi'\rangle_4|\vartheta\rangle_5|\vartheta'\rangle_6|x\rangle_7|y\rangle_8 \\ &+ |\varphi'\rangle_1|\varphi\rangle_2|\phi'\rangle_3|\phi\rangle_4|\vartheta'\rangle_5|\vartheta\rangle_6|y\rangle_7|x\rangle_8 \\ &- |\phi\rangle_1|\phi'\rangle_2|\vartheta\rangle_3|\vartheta'\rangle_4|x\rangle_5|y\rangle_6|\varphi\rangle_7|\varphi'\rangle_8 \\ &+ |\phi'\rangle_1|\phi\rangle_2|\vartheta'\rangle_3|\vartheta\rangle_4|y\rangle_5|x\rangle_6|\varphi'\rangle_7|\varphi\rangle_8 \\ &+ |\vartheta\rangle_1|\vartheta'\rangle_2|x\rangle_3|y\rangle_4|\varphi\rangle_5|\varphi'\rangle_6|\phi\rangle_7|\phi'\rangle_8 \\ &- |\vartheta'\rangle_1|\vartheta\rangle_2|y\rangle_3|x\rangle_4|\varphi'\rangle_5|\varphi\rangle_6|\phi'\rangle_7|\phi\rangle_8)\end{aligned} \tag{2.131}$$

2.13 Quantum entanglement probability amplitudes for $n = N = 3$

The triangular configuration applicable to the $n = N = 3$ case is illustrated in figure 2.7.

Here, the three quanta are shown as emitted at a relative angle of $2\pi/3$. The material presented in this chapter is a summary and extension of previous publications (Duarte 2015, 2016, 2019).

In this approach it is assumed that the three quanta are *indistinguishable*, and that $d_1 = d_2 = d_3 = d$. Under these circumstances the Dirac–Feynman principle leads to

$$\langle d|s\rangle = \langle d|p_3\rangle\langle p_3|s\rangle + \langle d|p_2\rangle\langle p_2|s\rangle + \langle d|p_1\rangle\langle p_1|s\rangle \tag{2.132}$$

and following the abstraction of $\langle d$

$$|D\rangle_3 = |p_3\rangle\langle p_3|s\rangle \tag{2.133}$$

$$|D\rangle_2 = |p_2\rangle\langle p_2|s\rangle \tag{2.134}$$

$$|D\rangle_1 = |p_1\rangle\langle p_1|s\rangle \tag{2.135}$$

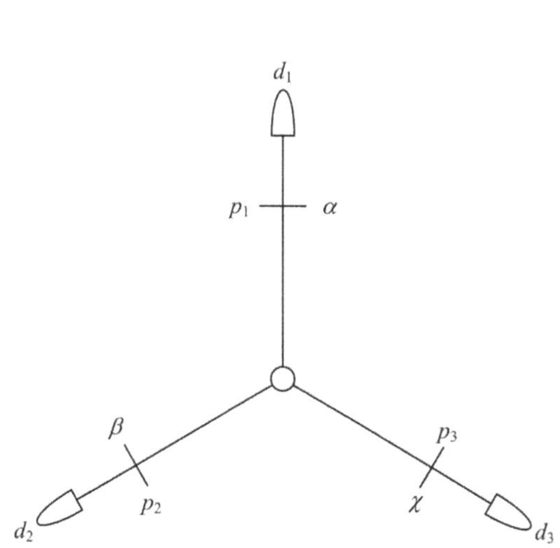

Figure 2.7. Schematics describing the experimental configuration to achieve quantum entanglement for $n = N = 3$ (Duarte 2019).

so that equation (2.132) becomes

$$|s\rangle = (|D\rangle_3 + |D\rangle_2 + |D\rangle_1) \tag{2.136}$$

Again, applying the Dirac identity $|X\rangle = |a\rangle_1|b\rangle_2|c\rangle_3 \dots |g\rangle_n$ (Dirac 1958), and using $|C\rangle_m$'s to differentiate from the pure interferometric case

$$|C\rangle_1 = |\alpha\rangle_1|\beta\rangle_2|\chi\rangle_3 \tag{2.137}$$

$$|C\rangle_2 = |\chi\rangle_1|\alpha\rangle_2|\beta\rangle_3 \tag{2.138}$$

$$|C\rangle_3 = |\beta\rangle_1|\chi\rangle_2|\alpha\rangle_3 \tag{2.139}$$

At this stage the principles established in equations (2.113)–(2.115) are applied with one caveat: *Hamilton's quaternions* (Hamilton 1866) are needed in order to satisfy the normalization equation

$$1 = ||\psi\rangle_I|^2 + ||\psi\rangle_{II}|^2 + ||\psi\rangle_{III}|^2 \tag{2.140}$$

The use of quaternions leads to

$$|\psi\rangle_I = 3^{-1/2}(|C\rangle_1 + |C\rangle_2 + i|C\rangle_3) \tag{2.141}$$

$$|\psi\rangle_{II} = 3^{-1/2}(|C\rangle_1 - |C\rangle_2 + i|C\rangle_3) \tag{2.142}$$

$$|\psi\rangle_{III} = 3^{-1/2}(i|C\rangle_1 + j|C\rangle_2 + k|C\rangle_3) \tag{2.143}$$

Substituting equations (2.137)–(2.139) into equations (2.141)–(2.143) leads to the explicit probability amplitudes

$$|\psi\rangle_I = 3^{-1/2}(|\alpha\rangle_1|\beta\rangle_2|\chi\rangle_3 + |\chi\rangle_1|\alpha\rangle_2|\beta\rangle_3 + i|\beta\rangle_1|\chi\rangle_2|\alpha\rangle_3) \tag{2.144}$$

$$|\psi\rangle_{II} = 3^{-1/2}(|\alpha\rangle_1|\beta\rangle_2|\chi\rangle_3 + |\chi\rangle_1|\alpha\rangle_2|\beta\rangle_3 - i|\beta\rangle_1|\chi\rangle_2|\alpha\rangle_3) \tag{2.145}$$

$$|\psi\rangle_I = 3^{-1/2}(i|\alpha\rangle_1|\beta\rangle_2|\chi\rangle_3 + j|\chi\rangle_1|\alpha\rangle_2|\beta\rangle_3 + k|\beta\rangle_1|\chi\rangle_2|\alpha\rangle_3) \tag{2.146}$$

2.14 Quantum entanglement probability amplitudes for $n = N = 6$

The hexagonal configuration applicable to the $n = N = 6$ case is illustrated in figure 2.8.

Here, the six quanta are shown as emitted at a relative angle of $\pi/3$. The material presented in this chapter is a summary and extension of previous publications (Duarte 2015, 2016, 2019).

In this approach it is assumed that the six quanta are *indistinguishable*, and that $d_1 = d_2 = d_3 = d_4 = d_5 = d_6 = d$. Under these circumstances the Dirac–Feynman principle leads to

$$|s\rangle = (|D\rangle_6 + |D\rangle_5 + |D\rangle_4 + |D\rangle_3 + |D\rangle_2 + |D\rangle_1) \tag{2.147}$$

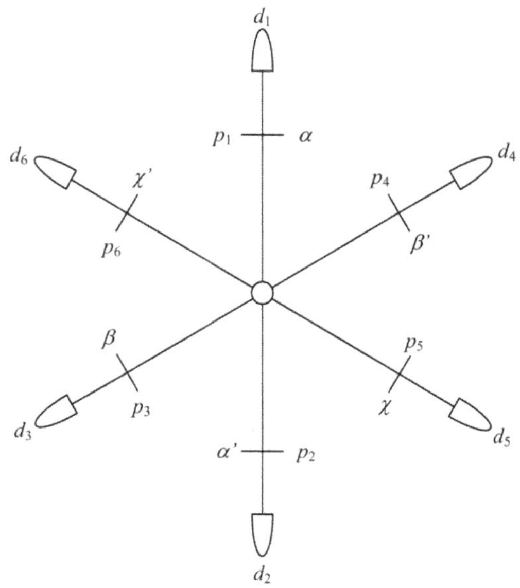

Figure 2.8. Schematics describing the experimental configuration to achieve quantum entanglement for $n = N = 6$ (Duarte 2019).

Observing figure 2.8 while labeling the polarization alternatives (α, α'), (β, β'), and (χ, χ'), the six combined states are

$$|C\rangle_1 = |\alpha\rangle_1|\alpha'\rangle_2|\beta\rangle_3|\beta'\rangle_4|\chi\rangle_5|\chi'\rangle_6 \tag{2.148}$$

$$|C\rangle_2 = |\alpha'\rangle_1|\alpha\rangle_2|\beta'\rangle_3|\beta\rangle_4|\chi'\rangle_5|\chi\rangle_6 \tag{2.149}$$

$$|C\rangle_3 = |\chi\rangle_1|\chi'\rangle_2|\alpha\rangle_3|\alpha'\rangle_4|\beta\rangle_5|\beta'\rangle_6 \tag{2.150}$$

$$|C\rangle_4 = |\chi'\rangle_1|\chi\rangle_2|\alpha'\rangle_3|\alpha\rangle_4|\beta'\rangle_5|\beta\rangle_6 \tag{2.151}$$

$$|C\rangle_5 = |\beta\rangle_1|\beta'\rangle_2|\chi\rangle_3|\chi'\rangle_4|\alpha\rangle_5|\alpha'\rangle_6 \tag{2.152}$$

$$|C\rangle_6 = |\beta'\rangle_1|\beta\rangle_2|\chi'\rangle_3|\chi\rangle_4|\alpha'\rangle_5|\alpha\rangle_6 \tag{2.153}$$

The quaternion version of the generalized quantum entanglement equation (2.113) can be expressed as (Duarte 2019)

$$|\psi\rangle_R = N^{-1/2}\sum_{j=1}^{N}(\pm)q|C\rangle_{N+1-j} \tag{2.154}$$

where q represents either 1 or a quaternion i, j, k needed to satisfy the normalization condition

$$1 = ||\psi\rangle_I|^2 + ||\psi\rangle_{II}|^2 + ||\psi\rangle_{III}|^2 + ||\psi\rangle_{IV}|^2 + ||\psi\rangle_V|^2 + ||\psi\rangle_{VI}|^2 \tag{2.155}$$

hence the six probability amplitudes are

$$|\psi\rangle_I = 6^{-1/2}(|C\rangle_1 + |C\rangle_2 + |C\rangle_3 + |C\rangle_4 + i|C\rangle_5 + i|C\rangle_6) \tag{2.156}$$

$$|\psi\rangle_{II} = 6^{-1/2}(|C\rangle_1 + |C\rangle_2 + |C\rangle_3 + |C\rangle_4 - i|C\rangle_5 - i|C\rangle_6) \tag{2.157}$$

$$|\psi\rangle_{III} = 6^{-1/2}(|C\rangle_1 + |C\rangle_2 - |C\rangle_3 - |C\rangle_4 + i|C\rangle_5 + i|C\rangle_6) \tag{2.158}$$

$$|\psi\rangle_{IV} = 6^{-1/2}(|C\rangle_1 + |C\rangle_2 - |C\rangle_3 - |C\rangle_4 - i|C\rangle_5 - i|C\rangle_6) \tag{2.159}$$

$$|\psi\rangle_V = 6^{-1/2}(i|C\rangle_1 + i|C\rangle_2 + j|C\rangle_3 + j|C\rangle_4 + k|C\rangle_5 + k|C\rangle_6) \tag{2.160}$$

$$|\psi\rangle_{VI} = 6^{-1/2}(-i|C\rangle_1 - i|C\rangle_2 - j|C\rangle_3 - j|C\rangle_4 - k|C\rangle_5 - k|C\rangle_6) \tag{2.161}$$

Substitution of the states (2.148)–(2.153) into equations (2.156)–(2.161) yields the explicit probability amplitudes as a function of (α, α'), (β, β'), and (χ, χ').

2.15 Beyond single quanta-pair quantum entanglement

Going back to the original Pryce-Ward probability amplitude on quantum entanglement which can be re-expressed as

$$|\psi\rangle_- = 2^{-1/2}(|x\rangle_I|y\rangle_{II} - |y\rangle_I|x\rangle_{II}) \tag{2.162}$$

there is nothing in Dirac's rules, for bosons, that would prevent states such as $|x\rangle_I$ and $|y\rangle_{II}$ from becoming combined states of indistinguishable polarized quanta ensembles such that

$$|x\rangle_I = |x\rangle_1|x\rangle_2|x\rangle_3\ldots|x\rangle_g\ldots \tag{2.163}$$

$$|y\rangle_{II} = |y\rangle_1|y\rangle_2|y\rangle_3\ldots|y\rangle_g\ldots \tag{2.164}$$

That is, all the indistinguishable quanta comprising ensemble I are polarized in the $|x\rangle$ state whilst all the indistinguishable quanta comprising ensemble II are polarized in the $|y\rangle$ state. Ultimately this means that ensembles $|x\rangle_I$ and $|y\rangle_{II}$ are themselves entangled. This opens the possibility of designing sources of *entangled laser emission* thus vastly improving the signal-to-noise conditions for quantum communications. The detection statistics of states with n indistinguishable bosons is discussed by Feynman *et al* (1965).

Tunable narrow-linewidth laser sources randomly emitting $|x\rangle_I$ and $|y\rangle_{II}$ states in opposite directions, via the use of intracavity elements driven by quantum random number generators (QRNG), have been suggested by Duarte (2018). This class of laser source has yet to be demonstrated experimentally.

2.16 Discussion

The Dirac–Feynman interferometric principle

$$\langle d|s\rangle = \sum_{j=1}^{N}\langle d|j\rangle\langle j|s\rangle$$

is a profound principle of nature. It describes beautifully and succinctly the interference of a single photon, the interference of populations of indistinguishable photons, and can be extended to even accurately describe the interference of semi-coherent radiation.

The Dirac–Feynman principle, in conjunction with the Dirac identity

$$|X\rangle = |a\rangle_1|b\rangle_2|c\rangle_3\ldots|g\rangle_n$$

provides a transparent and logical framework to derive the superposition probability amplitude of quantum entanglement. It does so in the absence of mysteries or paradoxes.

Problems

2.1 Using the Dirac–Feynman principle expressed in equation (2.2) write the probability amplitude applicable to an interferometer including three intra interferometric planes of N-slits separated by the spaces $D_{\langle k|j\rangle}$, $D_{\langle l|k\rangle}$, and $D_{\langle d|l\rangle}$.

2.2 Show that the quantum probability given in equation (2.5) can be expressed in more explicit notation as equation (2.12).

2.3 Show that the quantum probability given in equation (2.12) can be expressed in more explicit notation as equation (2.14).

2.4 Expand equation (2.20) to arrive to an explicit equation for $N = 6$.

2.5 Show that the dimensions for the quantum intensity, equation (2.26), are given in J s^{-1} m^{-2} or W m^{-2}.

2.6 Derive the quantum entanglement probability amplitudes for $n = N = 3$. Note that the solution to this problem involves Hamilton's quaternions.

2.7 Show that the generalized equations for the quantum entanglement probability amplitudes, equations (2.113)–(2.115), can lead to equations (2.86)–(2.89) for $n = N = 4$.

2.8 Use the generalized equations for the quantum entanglement probability amplitudes, equations (2.113)–(2.115), to derive explicit probability amplitudes, $|\psi\rangle_{II}$, $|\psi\rangle_{III}$, $|\psi\rangle_{IV}\ldots|\psi\rangle_{XVI}$, for $n = N = 2^4$.

2.9 Use the definition of quaternions to show that equation (2.140) leads to the probability amplitude equations (2.141)–(2.143).

References

Born M 1926 Zur quantenmechanik der stoßvorgänge *Z. Phys.* **37** 863–27

Dirac P A M 1939 A new notation for quantum mechanics *Math. Proc. Camb. Phil. Soc.* **35** 416–8

Dirac P A M 1958 *The Principles of Quantum Mechanics* 4th edn (Oxford: Oxford University Press)

Duarte F J 1991 Dispersive dye lasers *High Power Dye Lasers* ed F J Duarte (Berlin: Springer) ch 2

Duarte F J 1993 On a generalized interference equation and interferometric measurements *Opt. Comm.* **103** 8–14

Duarte F J 1995 Interferometric imaging *Tunable Laser Applications* ed F J Duarte (New York: Marcel Dekker) ch 5

Duarte F J 1999 Multiple-prism grating solid-state dye laser oscillator: optimized architecture *Appl. Opt.* **38** 6347–9

Duarte F J 2003 *Tunable Laser Optics* (New York: Elsevier Academic)

Duarte F J 2004 Comment on 'Reflection, refraction and multislit interference' *Eur. J. Phys.* **25** L57–8

Duarte F J 2013a The probability amplitude for entangled polarizations: an interferometric approach *J. Mod. Opt.* **60** 1585–7

Duarte F J 2013b Tunable laser optics: applications to optics and quantum optics *Prog. Quantum Electron.* **37** 326–47

Duarte F J 2014 *Quantum Optics for Engineers* (New York: CRC Press)

Duarte F J 2015 *Tunable Laser Optics* 2nd edn. (New York: CRC Press)

Duarte F J 2016 Secure space-to-space interferometric communications and its nexus to the physics of quantum entanglement *Appl. Phys. Rev.* **3** 041301

Duarte F J 2018 Organic lasers for *N*-channel quantum entanglement *Organic Lasers and Organic Photonics* ed F J Duarte (Bristol: IOP Publishing) ch 15

Duarte F J 2019 *Fundamentals of Quantum Entanglement* (Bristol: IOP Publishing)

Duarte F J and Paine D J 1989 Quantum mechanical description of *N*-slit interference phenomena *Proceedings of the International Conference on Lasers'88* ed R C Sze and F J Duarte (McLean, VA: STS Press) 552–4

Duarte F J and Taylor T S 2017 Quantum entanglement probability amplitudes in multiple channels: an interferometric approach *Optik* **139** 222–30

Duarte F J, Taylor T S, Black A M and Olivares I E 2013 Diffractive patters superimposed over propagating *N*-slit interferograms *J. Mod. Opt.* **60** 136–40

Feynman R P and Hibbs A R 1965 *Quantum Mechanics and Path Integrals* (New York: McGraw-Hill)

Feynman R P, Leighton R B and Sands M 1965 *The Feynman Lectures on Physics* vol III (Reading, MA: Addison-Wesley)

Hamilton W R 1866 *Elements of Quaternions* (London: Longman Green & Co.)

Lamb W E 1995 Anti-photon *Appl. Phys.* B **60** 77–84

Madrid J, Gine J and Chemisana D 2020 Quantum fluctuations and the *N*-slit interference *Int. J. Theor. Phys.* **60** 1–9

Pryce M L H and Ward J C 1947 Angular correlation effects with annihilation radiation *Nature* **160** 435

Ward J C 1949 *Some Properties of the Elementary Particles* (Oxford: Oxford University Press)

IOP Publishing

Quantum Entanglement Engineering and Applications

F J Duarte and T S Taylor

Chapter 3

Indistinguishability

The central role that the concept of indistinguishability plays in quantum interference, quantum entanglement, and quantum mechanics as a whole is discussed.

3.1 Introduction

The concept of *indistinguishability* is crucial to quantum mechanics. In this regard, it should be clarified that the idea of indistinguishability in quantum mechanics refers to an indistinguishability that is total, intrinsic, as in the case of indistinguishable quanta, or indistinguishable electrons, an indistinguishability that is inherent.

An early introduction on the concept of indistinguishability in quantum mechanics, can be found in Dirac's foundational book *The Principles of Quantum Mechanics* (Dirac 1958). For instance, this concept manifests itself implicitly in his discussion on interference and the word *indistinguishable* is stated explicitly in his discussion leading to the identity relevant to quantum entanglement. The word indistinguishable as related to interference is also explicitly utilized by Feynman *et al* (1965).

In this chapter indistinguishability is discussed within the framework of quantum interference, Dirac's identities, and quantum entanglement.

3.2 Indistinguishability in quantum interference

As already discussed in chapter 2, the Dirac–Feynman superposition probability amplitude

$$\langle d|s\rangle = \sum_{j=1}^{N} \langle d|j\rangle \ \langle j|s\rangle \tag{3.1}$$

is the interferometric principle essential to describe N-slit interference. The fit between theory and experiment is superbly natural. Via Born's rule (Born 1926) the probability amplitude (3.1) thus derived leads to the superposition probability equation (Duarte 1993, 1995)

doi:10.1088/978-0-7503-3407-5ch3

$$\langle d|s\rangle\langle d|s\rangle^* = \left(\sum_{j=1}^{N}\langle d|j\rangle\ \langle j|s\rangle\right)\left(\sum_{j=1}^{N}\langle d|j\rangle\ \langle j|s\rangle\right)^* \tag{3.2}$$

which can then be expressed in complex form as (Duarte and Paine 1989, Duarte 1991, 1993)

$$|\langle d|s\rangle|^2 = \sum_{j=1}^{N}\Psi(r_j)\sum_{m=1}^{N}\Psi(r_m)e^{i(\Omega_m-\Omega_j)} \tag{3.3}$$

or more explicitly as

$$|\langle d|s\rangle|^2 = \sum_{j=1}^{N}\Psi(r_j)^2 + 2\sum_{j=1}^{N}\Psi(r_j)\left(\sum_{m=j+1}^{N}\Psi(r_m)\cos(\Omega_m - \Omega_j)\right) \tag{3.4}$$

Equations (3.2)–(3.4) are equivalent and they apply either to single-photon propagation or to the propagation of an *ensemble of indistinguishable quanta* as available from narrow-linewidth lasers (Duarte 1993, 2003).

Measuring the interferometric probability might sound as removed from the practical sphere. What is not often realized is that measuring the intensity of a single photon, or the intensity distribution created by the interaction of an ensemble of indistinguishable quanta, is a quantum measurement. More specifically, the measured intensity for a single photon depends on the quantum probability as described by (Duarte 2014, 2019)

$$\mathscr{I}(\nu) = K\ \ h\nu\ \ \langle d|s\rangle\langle d|s\rangle^* \tag{3.5}$$

while for an ensemble of r indistinguishable photons the intensity is described by

$$\mathscr{I}(\nu) = r\ \ K\ \ h\nu|\langle d|s\rangle\langle d|s\rangle^*| \tag{3.6}$$

where K is a constant related to the intrinsic characteristics of the emission and has units of $m^{-2}\ s^{-1}$, $h = 6.626\ 070\ 15 \times 10^{-34}$ Js is Planck's constant, ν is the emission frequency in Hz, and $\langle d|s\rangle\langle d|s\rangle^*$ is the dimensionless probability. For a predetermined fixed frequency ν, the measured intensity is simply proportional to the quantum probability

$$I \propto\ |\langle d|s\rangle|^2 \tag{3.7}$$

In practice, the measured intensity faithfully represents the spatial distribution of the predicted quantum probability distribution (Duarte 1993, 1995). For further details on the intensity $I(\nu)$ the reader should refer to chapter 2.

3.3 Indistinguishability in Dirac's identities

As highlighted in previous publications (Duarte 2013a, 2013b, 2014, 2019), crucial to the derivation of the quantum entanglement probability amplitude (Pryce and Ward 1947)

$$|\psi\rangle = 2^{-1/2}(|x\rangle_1|y\rangle_2 - |y\rangle_1|x\rangle_2) \tag{3.8}$$

while utilizing the Dirac–Feynman interferometric principle (Dirac 1958, Feynman *et al* 1965)

$$\langle d|s\rangle = \sum_{j=1}^{N}\langle d|j\rangle\ \langle j|s\rangle$$

is the Dirac identity (Dirac 1958)

$$|X\rangle = |a\rangle_1|b\rangle_2|c\rangle_3\ldots|g\rangle_n \tag{3.9}$$

which means that indistinguishable quanta 1, 2, 3, … , n can be in different states. When referring to this 'curious phenomena… having no analogue in classical theory' Dirac is writing about quanta of the *'same kind'* and *'absolutely indistinguishable from one another'* even though in the notation they are labeled with distinct numbers (Dirac 1958). Dirac arrived at identity (3.9) by initially considering $|a\rangle_1, |b\rangle_1, |c\rangle_1\ldots|g\rangle_1$ for photon 1, $|a\rangle_2, |b\rangle_2, |c\rangle_2\ldots|g\rangle_2$ for photon 2, $|a\rangle_3, |b\rangle_3, |c\rangle_3\ldots|g\rangle_3$ for quanta 3, and so on.

For bosons, Dirac tabulates the various possible series of states in the matrix determinant (Dirac 1958)

$$\begin{vmatrix} |a\rangle_1 & |a\rangle_2 & |a\rangle_3 & \cdot & \cdot & \cdot & |a\rangle_n \\ |b\rangle_1 & |b\rangle_2 & |b\rangle_3 & \cdot & \cdot & \cdot & |b\rangle_n \\ |c\rangle_1 & |c\rangle_3 & |c\rangle_3 & \cdot & \cdot & \cdot & |c\rangle_n \\ \vdots & \vdots & \vdots & \vdots & \vdots & \vdots & \vdots \\ |g\rangle_1 & |g\rangle_2 & |g\rangle_3 & \cdot & \cdot & \cdot & |g\rangle_n \end{vmatrix} \tag{3.10}$$

In Dirac's writings there is nothing that prevents ensembles of indistinguishable quanta in an identical state of polarization. In this regard Dirac-type identities should be applicable, for instance, to describe single-transverse-mode single-longitudinal-mode coherent emission of indistinguishable quanta applicable for both, $|x\rangle$ and $|y\rangle$, states of polarization.

For instance, the ensemble for indistinguishable quanta in the $|x\rangle$ polarization state can be written as

$$|\alpha\rangle = |x\rangle_1|x\rangle_2|x\rangle_3\ldots|x\rangle_n \tag{3.11}$$

and for indistinguishable quanta in the $|y\rangle$ polarization state

$$|\beta\rangle = |y\rangle_1|y\rangle_2|y\rangle_3\ldots|y\rangle_n \tag{3.12}$$

These identities describe precisely and elegantly what coherent emission is at a very fundamental level: ensembles of indistinguishable quanta in a given state of polarization. This perspective is in line with Dirac's assertion that '*with Bose statistics the probability of two particles being in the same state is greater than with classical statistics*' (Dirac 1958). In the case of optical quantum entanglement the word particles should be replaced by quanta. The detection statistics of states with n

insistinguishable bosons, in other words n insistinguishable quanta, is discussed by Feynman *et al* (1965).

Understanding this concept allows us to provide a transparent explanation of highly coherent emission from electrically-excited organic semiconductors integrated within interferometric configurations (Duarte *et al* 2005, Duarte and Vaeth 2018). Further details are given in appendix A.

Note: Dirac does not provide a derivation for the determinant (3.11). However, an array of $n \times n$ states can be arrived at by considering combined states of the form

$$|Y\rangle = |a\rangle_1|b\rangle_1|c\rangle_1 \ldots |g\rangle_1 \tag{3.13}$$

$$|Z\rangle = |a\rangle_2|b\rangle_2|c\rangle_2 \ldots |g\rangle_2 \tag{3.14}$$

$$|\xi\rangle = |a\rangle_3|b\rangle_3|c\rangle_3 \ldots |g\rangle_3 \tag{3.15}$$

$$|\zeta\rangle = |a\rangle_n|b\rangle_n|c\rangle_n \ldots |g\rangle_n \tag{3.16}$$

The transpose of this array can be written as

$$\begin{matrix} |a\rangle_1 & |a\rangle_2 & |a\rangle_3 & \ldots & |a\rangle_n \\ |b\rangle_1 & |b\rangle_2 & |b\rangle_3 & \ldots & |b\rangle_n \\ |c\rangle_1 & |c\rangle_3 & |c\rangle_3 & \ldots & |c\rangle_n \\ \vdots & \vdots & \vdots & \vdots & \vdots \\ |g\rangle_1 & |g\rangle_2 & |g\rangle_3 & \ldots & |g\rangle_n \end{matrix} \tag{3.17}$$

which has the same form as Dirac's determinant. The rows describe quanta in the same state and the diagonal is the Dirac identity

$$|X\rangle = |a\rangle_1|b\rangle_2|c\rangle_3 \ldots |g\rangle_n$$

A relevant Dirac statement in reference to quanta indistinguishability and the $n \times n$ state array given in (3.10) or (3.17) is 'one cannot say which particle is in which state, each *quantum* being equally likely to be in any state' and he adds that for bosons 'two or more particles can be in the same state '(Dirac 1958). In the reference to bosons, and the context of this book, the words particle and particles should be replaced by quantum and quanta, respectively.

3.4 Indistinguishability in quantum entanglement

The derivation of the probability amplitude of quantum entanglement within an interferometric framework is discussed in chapter 6. Here, emphasis is given to aspects of this derivation relevant to the phenomenon of indistinguishability.

The experimental diagram applicable to linear quantum entanglement including two quanta propagating in two channels, that is, $n = N = 2$, in opposite directions is depicted in figure 3.1. Here, s is a photon source of ν_1 and ν_2 emitted in the $+z$- and $-z$-directions, propagating through the polarization analyzers p_1 and p_2 toward the detectors d_1 and d_2, respectively.

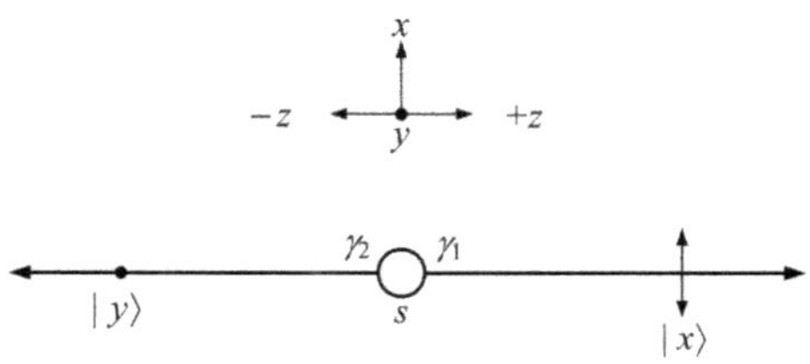

Figure 3.1. Experimental diagram applicable to linear quantum entanglement involving two quanta propagating in two channels, that is, $n = N = 2$, in opposite directions, from Duarte (2019).

Assuming that the quanta emitted by the source are *indistinguishable*, that is $\nu_1 = \nu_2 = \nu$, then applying the Dirac interferometric principle

$$\langle d|s\rangle = \sum_{j=1}^{N} \langle d|p_j\rangle\langle p_j|s\rangle \tag{3.18}$$

leads to the expansion

$$\langle d|s\rangle = \langle d|p_2\rangle\langle p_2|s\rangle + \langle d|p_1\rangle\langle p_1|s\rangle \tag{3.19}$$

Implicitly, this approach requires that $d_1 = d_2 = d$, which is quite reasonable for a pair of matched detectors. Abstracting d from equation (3.19), this expression can be re stated as

$$|s\rangle = |p_1\rangle\langle p_2|s\rangle + |p_1\rangle\langle p_1|s\rangle \tag{3.20}$$

Next, using the Dirac identity $|\psi\rangle = |j\rangle\langle j|\psi\rangle$ allows for the expressions

$$|D\rangle_2 = |p_2\rangle\langle p_2|s\rangle \tag{3.21}$$

and

$$|D\rangle_1 = |p_1\rangle\langle p_1|s\rangle \tag{3.22}$$

Substituting identities (3.21) and (3.22) into equation (3.20) yields

$$|s\rangle = (|D\rangle_2 + |D\rangle_1) \tag{3.23}$$

Equation (3.18) is a superposition probability amplitude $|s\rangle$ representing the linear combination of two probability amplitudes, $|D\rangle_1$ and $|D\rangle_2$, and applies equally to an interferometric situation involving a single photon, or an ensemble of indistinguishable photons.

Equation (3.23) also applies in the quantum entanglement framework. The crucial innovation occurs at the very stage that invokes the use of the Dirac identity for *similar* quanta, which is synonymous with *indistinguishable quanta.* This Dirac identity indicates that a single state $|X\rangle$ can be comprised by the multiplication of a series of different states while each of these states is associated with a particular quantum

$$|X\rangle = |a\rangle_1|b\rangle_2|c\rangle_3\ldots|g\rangle_n$$

This very step introduces a divergence from the physics of a purely interferometric indistinguishable situation given that the possibility of separate quanta, quantum 1 and quantum 2, is allowed with each quanta in a different alternative state. That is, state $|x\rangle$ and state $|y\rangle$. Rewriting $|D\rangle_1$ and $|D\rangle_2$ in terms of two polarization states as per identity (3.9), and using $|C\rangle_1$ and $|C\rangle_2$ to distinguish from a pure interferometric event, leads to

$$|s\rangle = (|C\rangle_2 + |C\rangle_1) \tag{3.24}$$

where

$$|C\rangle_2 = |x\rangle_1|y\rangle_2 \tag{3.25}$$

and

$$|C\rangle_1 = |y\rangle_1|x\rangle_2 \tag{3.26}$$

Inserting identities (3.25) and (3.26) in the expression for the probability amplitude in equation (3.24) leads directly to (Duarte 2013a, 2013b, 2014, 2016)

$$|s\rangle = (|x\rangle_1|y\rangle_2 + |y\rangle_1|x\rangle_2) \tag{3.27}$$

Following normalization, and designating $|s\rangle$ as $|\psi\rangle$, equation (3.22) can be expressed as

$$|\psi\rangle_+ = 2^{-1/2}(|x\rangle_1|y\rangle_2 + |y\rangle_1|x\rangle_2) \tag{3.28}$$

and its linear combination is

$$|\psi\rangle_- = 2^{-1/2}(|x\rangle_1|y\rangle_2 - |y\rangle_1|x\rangle_2) \tag{3.29}$$

which is the iconic probability amplitude for quantum entanglement (Pryce and Ward 1947, Ward 1949).

Note: when the quanta are emitted they are totally indistinguishable. The quanta continue to be indistinguishable until the measurement is performed. At the measurement the superposition probability amplitude $|\psi\rangle_-$ becomes the superposition probability

$$|\psi\rangle_-|\psi\rangle_-^* = P(\phi_1, \phi_2) = -\cos 2(\phi_1 - \phi_2) \tag{3.30}$$

and the opportunity to differentiate quantum 1 from quantum 2 vanishes. Only post measurement, once entirely in the classical domain, is it possible to assign one polarization to photon 1 and the orthogonal polarization to photon 2, or vice versa.

3.5 Indistinguishability in quanta ensembles

The superposition probability amplitudes for quantum entanglement, $|\psi\rangle_+$ and $|\psi\rangle_-$, can be re expressed as

$$|\psi\rangle_+ = 2^{-1/2}(|x\rangle_I|y\rangle_{II} + |y\rangle_I|x\rangle_{II}) \tag{3.31}$$

$$|\psi\rangle_- = 2^{-1/2}(|x\rangle_I|y\rangle_{II} - |y\rangle_I|x\rangle_{II}) \tag{3.32}$$

There is nothing in Dirac's rules for bosons that would prevent states such as $|x\rangle_I$ and $|y\rangle_{II}$ from being combined states of indistinguishable polarized quanta ensembles such that

$$|x\rangle_I = |x\rangle_1|x\rangle_2|x\rangle_3\ldots|x\rangle_g\ldots \tag{3.33}$$

$$|y\rangle_{II} = |y\rangle_1|y\rangle_2|y\rangle_3\ldots|y\rangle_g\ldots \tag{3.34}$$

That is, all the indistinguishable quanta comprising ensemble I are polarized in the $|x\rangle$ state whilst all the indistinguishable quanta comprising ensemble II are polarized in the $|y\rangle$ state. Ultimately this means that ensembles $|x\rangle_I$ and $|y\rangle_{II}$ are themselves entangled. This perspective opens the possibility of designing sources of *entangled laser emission* thus vastly improving the signal-to-noise conditions for quantum communications (Duarte 2018).

3.6 Discussion

The concept of indistinguishability is inherent to the interferometric Dirac–Feynman principle

$$\langle d|s\rangle = \sum_{j=1}^{N}\langle d|j\rangle\ \langle j|s\rangle$$

The Dirac identity $|X\rangle = |a\rangle_1|b\rangle_2|c\rangle_3\ldots|g\rangle_n$ refers to a single state $|X\rangle$ composed by an assembly of different states $|a\rangle|b\rangle|c\rangle\ldots|g\rangle$ each of them associated with an individual quantum from the series of undistinguishable quanta 1, 2, 3,..., n. The transition from a purely interferometric situation to a quantum entanglement condition occurs once single whole states such as $|C\rangle_1$ and $|C\rangle_2$ are allowed to be identified as product states such as $|x\rangle_1|y\rangle_2$ and $|y\rangle_1|x\rangle_2$. However, this occurs in Dyson's first layer out of reach from the experimenter or measurer (see chapters 8 and 10).

As a final thought it can be argued that whenever an experimental situation involving indistinguishability arises, the physics has to be treated quantum mechanically. In this regard, *indistinguishability is quantum*. However, quantum mechanics also includes further physical phenomena such as quantum entanglement, which allow, post measurement, discrimination between different sates, such as $|x\rangle$ and $|y\rangle$, in quanta that were indistinguishable at their origin.

In the introduction, reference was made to the concept of intrinsic or inherent indistinguishability as in ensembles of identical quanta or electrons. It remains to be seen if quantum mechanical laws also apply to broader definitions of indistinguishability. For example, ensembles integrated by slightly different atomic elements, which in principle could be differentiated, that by choice are treated as ensembles comprised of indistinguishable elements.

Problems

3.1 Show that equation (3.4) follows from equation (3.3).

3.2 Using the identities of the type given in equations (3.11) and (3.12), arrive at the other two superposition quantum entanglement probability amplitudes allowed for $n = N = 2$.
3.3 For $n = N = 2$ discuss what would happen to the superposition probability amplitude for quantum entanglement if the two detectors were different, that is $d_1 \neq d_2$.

References

Born M 1926 Zur quantenmechanik der stoßvorgänge *Z. Phys.* **37** 863–27

Dirac P A M 1958 *The Principles of Quantum Mechanics* (Oxford: Oxford University Press)

Duarte F J 1991 Dispersive dye lasers *High Power Dye Lasers* ed F J Duarte (Berlin: Springer) ch 2

Duarte F J 1993 On a generalized interference equation and interferometric measurements *Opt. Comm.* **103** 8–14

Duarte F J 1995 Interferometric imaging *Tunable Laser Applications* ed F J Duarte (New York: Marcel Dekker) ch 5

Duarte F J 2003 *Tunable Laser Optics* 1st edn (New York: Elsevier Academic))

Duarte F J 2013a The probability amplitude for entangled polarizations: an interferometric approach *J. Mod. Opt.* **60** 1585–7

Duarte F J 2013b Tunable laser optics: applications to optics and quantum optics *Prog. Quantum Electron.* **37** 326–47

Duarte F J 2014 *Quantum Optics for Engineers* (New York: CRC Press)

Duarte F J 2016 Secure space-to-space interferometric communications and its nexus to the physics of quantum entanglement *Appl. Phys. Rev.* **3** 041301

Duarte F J 2018 Organic lasers for *N*-channel quantum entanglement *Organic Lasers and Organic Photonics* ed F J Duarte (Bristol: IOP Publishing) ch 15

Duarte F J 2019 *Fundamentals of Quantum Entanglement* (Bristol: IOP Publishing)

Duarte F J, Liao L S and Vaeth K M 2005 Coherence characteristics of electrically excited tandem organic light-emitting diodes *Opt. Lett.* **30** 3072–4

Duarte F J and Paine D J 1989 Quantum mechanical description of *N*-slit interference phenomena *Proceedings of the International Conference on Lasers'88* ed R C Sze and F J Duarte (McLean, VA: STS Press) 552–4

Duarte F J and Vaeth K M 2018 Electrically-pumped organic semiconductor laser emission *Organic Lasers and Organic Photonics* ed F J Duarte (Bristol: IOP Publishing) ch 11

Feynman R P, Leighton R B and Sands M 1965 *The Feynman Lectures on Physics* vol III (Reading, MA: Addison-Wesley)

Pryce M L H and Ward J C 1947 Angular correlation effects with annihilation radiation *Nature* **160** 435

Ward J C 1949 *Some Properties of the Elementary Particles* (Oxford: Oxford University Press)

IOP Publishing

Quantum Entanglement Engineering and Applications

F J Duarte and T S Taylor

Chapter 4

Quantum interferometry via Dirac's *bra ket* notation

The probability amplitudes applicable to the N-slit interferometer and various beam-splitter based interferometers, such as the Mach–Zehnder interferometer, the Michelson interferometer, and the Sagnac interferometer are derived. The attention then focuses on the derivation of the matrix for the Hadamard gate and on interferometers based on a single-beam splitter.

4.1 Introduction

In this chapter, the physics of the N-slit interferometer is reviewed and the probability amplitudes for beam-splitter based interferometers such as the Mach–Zehnder interferometer, the Michelson interferometer, and the Sagnac interferometer are derived from first principles using Dirac's *bra ket* notation (Dirac 1939, 1958). For classical discussions on these interferometers the reader is referred to Michelson (1927) and Steel (1976).

All the probability amplitudes, also known as quantum states, derived in this chapter are based in the Dirac–Feynman interferometric principle (Dirac 1958, Feynman *et al* 1965).

The interferometric treatment given in this chapter utilizes the notation and style introduced by Duarte (2003, 2014) and, in part, is an improved extension of the review given by Duarte (2019).

4.2 The N-slit interferometer

As discussed in chapter 2 the N-slit interferometer, illustrated in figure 4.1, is described directly by the Dirac–Feynman *probability amplitude*

$$\langle d|s\rangle = \sum_{j=1}^{N} \langle d|j\rangle\langle j|s\rangle \tag{4.1}$$

doi:10.1088/978-0-7503-3407-5ch4

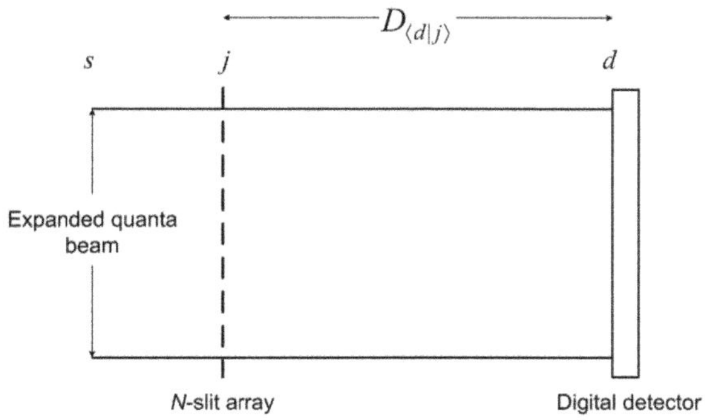

Figure 4.1. Schematics of the N-slit interferometer. The source is s, the N-slit array is j, and the interferometric plane, or digital detector is d. The intra interferometric distance from j to d is $D_{\langle d|j\rangle}$.

This probability amplitude, multiplied by its complex conjugate, as required by Born's rule (Born 1926), yields the one-dimensional interferometric *probability*

$$\langle d|s\rangle\langle d|s\rangle^* = \left(\sum_{j=1}^{N}\langle d|j\rangle\langle j|s\rangle\right)\left(\sum_{j=1}^{N}\langle d|j\rangle\langle j|s\rangle\right)^* \tag{4.2}$$

that can be expressed as (Duarte 1993, 2003, 2014)

$$\langle d|s\rangle\langle d|s\rangle^* = \sum_{j=1}^{N}\Psi(r_j)\sum_{m=1}^{N}\Psi(r_m)e^{i(\Omega_m-\Omega_j)} \tag{4.3}$$

$$\langle d|s\rangle\langle d|s\rangle^* = \sum_{j=1}^{N}\Psi(r_j)^2 + 2\sum_{j=1}^{N}\Psi(r_j)\left(\sum_{m=j+1}^{N}\Psi(r_m)\cos(\Omega_m - \Omega_j)\right) \tag{4.4}$$

These three equivalent probability equations apply to single-photon propagation or to the propagation of ensembles of indistinguishable photons Duarte (1993). Explicit long-hand versions of equation (4.36), for $N = 2, 3, \ldots, 5$, are given by Duarte (2014, 2015).

For instance, for $N = 5$ (Duarte 2014)

$$\begin{aligned}
|\langle x|s\rangle|^2 &= \Psi(r_1)^2 + \Psi(r_2)^2 + \Psi(r_3)^2 + \Psi(r_4)^2 + \Psi(r_5)^2 \\
&+ 2(\Psi(r_1)\Psi(r_2)\cos(\Omega_2 - \Omega_1) + \Psi(r_1)\Psi(r_3)\cos(\Omega_3 - \Omega_1) \\
&+ \Psi(r_1)\Psi(r_4)\cos(\Omega_4 - \Omega_1) \\
&+ \Psi(r_1)\Psi(r_5)\cos(\Omega_5 - \Omega_1) + \Psi(r_2)\Psi(r_3)\cos(\Omega_3 - \Omega_2) \\
&+ \Psi(r_2)\Psi(r_4)\cos(\Omega_4 - \Omega_2)+ \\
&+ \Psi(r_2)\Psi(r_5)\cos(\Omega_5 - \Omega_2) + \Psi(r_3)\Psi(r_4)\cos(\Omega_4 - \Omega_3) \\
&+ \Psi(r_3)\Psi(r_5)\cos(\Omega_5 - \Omega_3) \\
&+ \Psi(r_4)\Psi(r_5)\cos(\Omega_5 - \Omega_4))
\end{aligned} \tag{4.5}$$

and for $N = 7$(Duarte 2015)

$$\begin{aligned}
|\langle x|s\rangle|^2 &= \Psi(r_1)^2 + \Psi(r_2)^2 + \Psi(r_3)^2 + \Psi(r_4)^2 + \Psi(r_5)^2 + \Psi(r_6)^2 + \Psi(r_7)^2 \\
&+ 2(\Psi(r_1)\Psi(r_2)\cos(\Omega_2 - \Omega_1) + \Psi(r_1)\Psi(r_3)\cos(\Omega_3 - \Omega_1) + \Psi(r_1)\Psi(r_4)\cos(\Omega_4 - \Omega_1) \\
&+ \Psi(r_1)\Psi(r_5)\cos(\Omega_5 - \Omega_1) + \Psi(r_1)\Psi(r_6)\cos(\Omega_6 - \Omega_1) + \Psi(r_1)\Psi(r_7)\cos(\Omega_7 - \Omega_1) \\
&+ \Psi(r_2)\Psi(r_3)\cos(\Omega_3 - \Omega_2) + \Psi(r_2)\Psi(r_4)\cos(\Omega_4 - \Omega_2) + \Psi(r_2)\Psi(r_5)\cos(\Omega_5 - \Omega_2) \\
&+ \Psi(r_2)\Psi(r_6)\cos(\Omega_6 - \Omega_2) + \Psi(r_2)\Psi(r_7)\cos(\Omega_7 - \Omega_2) + \Psi(r_3)\Psi(r_4)\cos(\Omega_4 - \Omega_3) \\
&+ \Psi(r_3)\Psi(r_5)\cos(\Omega_5 - \Omega_3) + \Psi(r_3)\Psi(r_6)\cos(\Omega_6 - \Omega_3) + \Psi(r_3)\Psi(r_7)\cos(\Omega_7 - \Omega_3) \\
&+ \Psi(r_4)\Psi(r_5)\cos(\Omega_5 - \Omega_4) + \Psi(r_4)\Psi(r_6)\cos(\Omega_6 - \Omega_4) + \Psi(r_4)\Psi(r_7)\cos(\Omega_7 - \Omega_4) \\
&+ \Psi(r_5)\Psi(r_6)\cos(\Omega_6 - \Omega_5) + \Psi(r_5)\Psi(r_7)\cos(\Omega_7 - \Omega_5) + \Psi(r_6)\Psi(r_7)\cos(\Omega_7 - \Omega_6))
\end{aligned} \tag{4.6}$$

4.3 Interferometers configured by beam splitters

In this section Dirac's *bra ket* notation is applied to describe the single-beam splitter, the Mach–Zehnder interferometer, the Michelson interferometer, and the Sagnac interferometer.

4.3.1 The single-beam splitter

For the special case of a straightforward beam splitter illuminated by a single beam of light, as described in figure 4.2, in which detectors d and d' are identical, the probability amplitude is given by

$$\langle d|s\rangle = \langle d|j'\rangle\langle j'|s\rangle + \langle d|j\rangle\langle j|s\rangle \tag{4.7}$$

where j represents reflection at the beam splitter and j' stands for transmission.

Abstracting $\langle d$ and setting

$$|A\rangle = |j'\rangle\langle j'|s\rangle \tag{4.8}$$

$$|B\rangle = |j\rangle\langle j|s\rangle \tag{4.9}$$

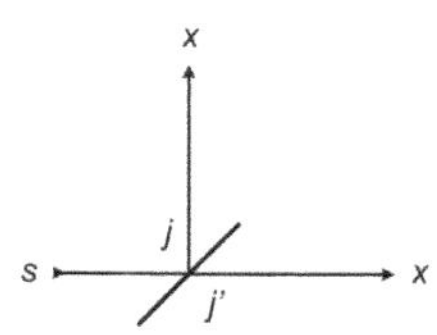

Figure 4.2. Ideal beam splitter under single-beam illumination. The incident beam is assumed of consisting either of a single photon or an ensemble of indistinguishable photons. Interaction of the photon with the beam splitter in the reflection mode is labeled as j whilst interaction of the photon with beam splitter in the transmission mode is assigned as j' (from Duarte 2019).

Equation (4.7) leads directly to

$$|s\rangle = (|A\rangle + |B\rangle) \tag{4.10}$$

Taking into account the linear combination of equation (4.7) and following normalization the probability amplitude takes the final form of

$$|s\rangle = 2^{-1/2}(|A\rangle \pm |B\rangle) \tag{4.11}$$

provided the two photons are detected by identical detectors ($d = d'$) situated at identical distances from the beam splitter. Note that if the detectors are not identical the experimental situation is represented by two separate probability amplitudes

$$\langle d|s\rangle = \langle d|j\rangle\langle j|s\rangle \tag{4.12}$$

$$\langle d'|s\rangle = \langle d'|j'\rangle\langle j'|s\rangle \tag{4.13}$$

in a non-superposition configuration.

4.3.2 The Mach–Zehnder interferometer

The Mach–Zehnder interferometer is comprised of an input beam splitter, an output beam splitter, and two mirrors M_1 and M_2, as depicted in figure 4.3. This interferometer provides two interferometric outputs, one at x and the other at x'. The interference of the counter propagating beams is described by the probability amplitude (Duarte 2003, 2014)

$$\langle d|s\rangle = \langle d|k'\rangle\langle k'|M_1\rangle\langle M_1|j\rangle\langle j|s\rangle + \langle d|k\rangle\langle k|M_2\rangle\langle M_2|j'\rangle\langle j'|s\rangle \tag{4.14}$$

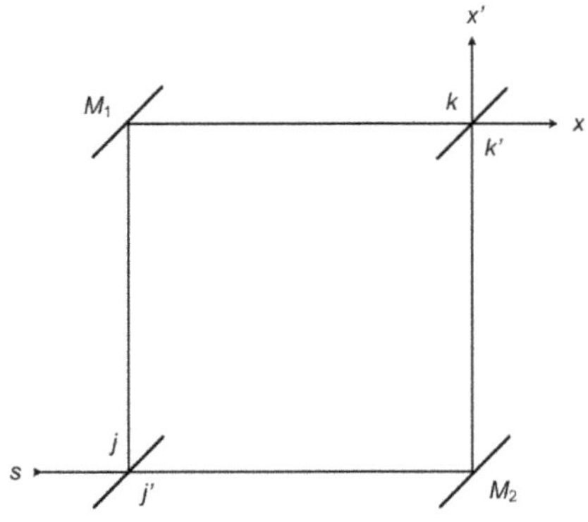

Figure 4.3. Schematics for the Mach–Zehnder interferometer. Mirrors M_1 and M_2 are assumed to be lossless and perfect. The beam splitters j and k are also assumed to be ideal, lossless, 50–50 partial reflectors. Interaction of the photon with the beam splitters in the reflection mode is labeled as j and k whilst interaction of the photon with beam splitter in the transmission mode is assigned as j' and k' (from Duarte 2019).

where j and k refer to the beam splitters in the reflective mode while j' and k' refer to the beam splitters in the transmission mode. Assuming mirrors M_1 and M_2 offer perfect reflectivity, equation (4.14) reduces to (Duarte 2003, 2014)

$$\langle d|s\rangle = \langle d|k'\rangle\langle k'|j\rangle\langle j|s\rangle + \langle d|k\rangle\langle k|j'\rangle\langle j'|s\rangle \tag{4.15}$$

Using the Dirac identity $|\phi\rangle = |j\rangle\langle j|\phi\rangle$, and abstracting, equation (4.15) reduces to

$$\langle d|s\rangle = \langle d|k'\rangle\langle k'|A\rangle + \langle d|k\rangle\langle k|B\rangle \tag{4.16}$$

where

$$|A\rangle = |j\rangle\langle j|s\rangle \tag{4.17}$$

and

$$|B\rangle = |j'\rangle\langle j'|s\rangle \tag{4.18}$$

Abstracting the $\langle d$, equation (4.16) can be expressed as

$$|s\rangle = (|k'\rangle\langle k'|A\rangle + |k\rangle\langle k|B\rangle) \tag{4.19}$$

Including the linear combination of equation (4.19), and following normalization, the overall probability amplitude becomes

$$|s\rangle = 2^{-1/2}(|C\rangle \pm |D\rangle) \tag{4.20}$$

where

$$|C\rangle = |k'\rangle\langle k'|A\rangle \tag{4.21}$$

and

$$|D\rangle = |k\rangle\langle k|B\rangle \tag{4.22}$$

For the d' detector

$$\langle d'|s\rangle = \langle d'|k\rangle\langle k|M_1\rangle\langle M_1|j\rangle\langle j|s\rangle + \langle d'|k'\rangle\langle k'|M_2\rangle\langle M_2|j'\rangle\langle j'|s\rangle \tag{4.23}$$

$$\langle d'|s\rangle = \langle d'|k\rangle\langle k|j\rangle\langle j|s\rangle + \langle d'|k'\rangle\langle k'|j'\rangle\langle j'|s\rangle \tag{4.24}$$

$$\langle d'|s\rangle = \langle d'|k\rangle\langle k|C\rangle + \langle d'|k'\rangle\langle k'|D\rangle \tag{4.25}$$

which also leads to an equation of the form of equation (4.20).

Equation (4.20) is the probability amplitude applicable to single-photon propagation, or the propagation of an ensemble of indistinguishable photons, in a Mach–Zehnder interferometer.

Mathematically speaking, it should also be noted that equations of the form of equation (4.16) are totally compatible with probability amplitudes constructed from the Dirac–Feynman principle for $N = 2$, that is,

$$\langle d|s\rangle = \sum_{j=1}^{N=2} \langle d|j\rangle\langle j|s\rangle \tag{4.26}$$

which is directly applicable to the double-slit interferometer (see chapter 2).

However, from a physics perspective, a Mach–Zehnder interferometer is different to a double-slit interferometer. In the double-slit interferometer the single photon, or the ensemble of indistinguishable photons, undergoes violent diffraction at the slits. This diffraction makes the physics between the two interferometers quite different. The only similarity between the two interferometers is that they are both two-path interferometers, that is, $N = 2$. However, while the Mach–Zehnder interferometer is a *two-beam* interferometer the double-slit, or two-slit, or Young, interferometer is a *parallel diffraction* interferometer (Duarte 2019).

4.3.3 The Michelson interferometer

The Michelson interferometer (Michelson 1927) is configured by one beam slitter and two mirrors M_1 and M_2 in an L-type geometry, as illustrated in figure 4.4. From an interferometric perspective, the beam splitter is illuminated by two inputs from which a single output emerges towards the detector d.

The probability amplitude describing single-photon propagation from the source s to the detector d is given by

$$\langle d|s\rangle = \langle d|j\rangle\langle j|M_2\rangle\langle M_2|j'\rangle\langle j'|s\rangle + \langle d|j'\rangle\langle j'|M_1\rangle\langle M_1|j\rangle\langle j|s\rangle \tag{4.27}$$

where j represents reflection at the beam splitter and j' transmission. For perfect mirror reflectivity equation (4.27) reduces to

$$\langle d|s\rangle = \langle d|j\rangle\langle j|j'\rangle\langle j'|s\rangle + \langle d|j'\rangle\langle j'|j\rangle\langle j|s\rangle \tag{4.28}$$

Abstraction, using the Dirac identity $|\phi\rangle = |j\rangle\langle j|\phi\rangle$, yields

$$\langle d|s\rangle = \langle d|j\rangle\langle j|s\rangle + \langle d|j'\rangle\langle j'|s\rangle \tag{4.29}$$

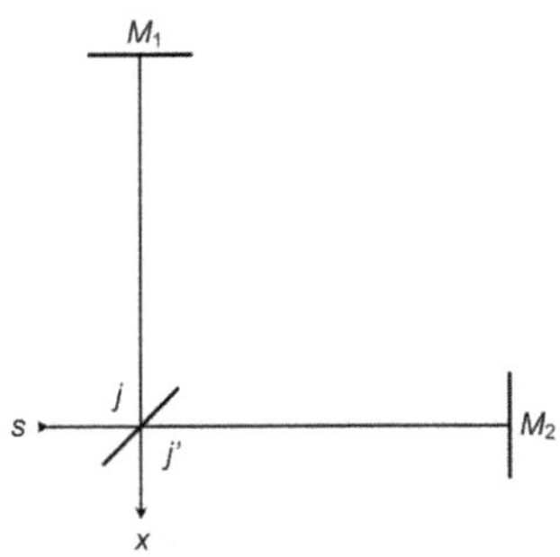

Figure 4.4. Michelson interferometer. See text for details (from Duarte 2019).

which leads directly to a probability amplitude of the form of equation (4.26) for $N = 2$.

Defining

$$|E\rangle = |j'\rangle\langle j'|s\rangle \tag{4.30}$$

and

$$|F\rangle = |j\rangle\langle j|s\rangle \tag{4.31}$$

Equation (4.29) becomes

$$|s\rangle = 2^{-1/2}(|E\rangle \pm |F\rangle) \tag{4.32}$$

which has the same form of the interferometric probability amplitude described by equation (4.20). This reinforces the observation that all the physics is performed at the beam splitter since the mirrors M_1 and M_2 are assumed to be perfect mirrors.

4.3.4 The Sagnac interferometer

The Sagnac interferometer is configured by one beam splitter and three mirrors M_1, M_2, and M_3 as shown in figure 4.5. Using the same definition for j and j', as previously, the probability amplitude is given by

$$\begin{aligned}\langle d|s\rangle &= \langle d|j\rangle\langle j|M_3\rangle\langle M_3|M_2\rangle\langle M_2|M_1\rangle\langle M_1|j\rangle\langle j|s\rangle \\ &+ \langle d|j'\rangle\langle j'|M_1\rangle\langle M_1|M_2\rangle\langle M_2|M_3\rangle\langle M_3|j'\rangle\langle j'|s\rangle\end{aligned} \tag{4.33}$$

For ideal lossless mirrors $M_1 = M_2 = M_3$

$$\langle j|M_3\rangle\langle M_3|M_2\rangle\langle M_2|M_1\rangle\langle M_1|j\rangle = 1 \tag{4.34}$$

$$\langle j'|M_1\rangle\langle M_1|M_2\rangle\langle M_2|M_3\rangle\langle M_3|j'\rangle = 1 \tag{4.35}$$

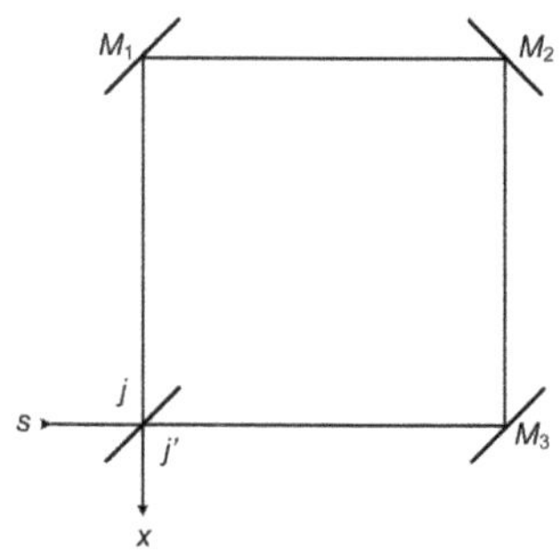

Figure 4.5. Sagnac interferometer with three mirrors M_1, M_2, and M_3. See text for details (from Duarte 2019).

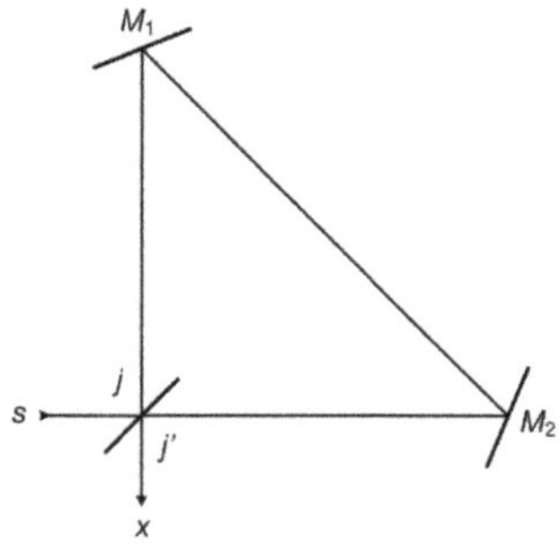

Figure 4.6. Triangular Sagnac interferometer incorporating mirrors M_1 and M_2. See text for details (from Duarte 2019).

and equation (4.33) can be expressed as

$$\langle d|s\rangle = \langle d|j\rangle\langle j|s\rangle + \langle d|j'\rangle\langle j'|s\rangle \tag{4.36}$$

Defining

$$|G\rangle = |j'\rangle\langle j'|s\rangle \tag{4.37}$$

$$|H\rangle = |j\rangle\langle j|s\rangle \tag{4.38}$$

Equation (4.36) becomes

$$|s\rangle = 2^{-1/2}(|G\rangle \pm |H\rangle) \tag{4.39}$$

Again, it should be noted that the physics of equation (4.39) can be traced back to the Dirac–Feynman probability amplitude for $N = 2$.

For the triangular Sagnac interferometer, including mirrors M_1 and M_2, as depicted in figure 4.6, the superposition probability amplitude becomes

$$\begin{aligned}\langle d|s\rangle &= \langle d|j\rangle\langle j|M_2\rangle\langle M_2|M_1\rangle\langle M_1|j\rangle\langle j|s\rangle \\ &+ \langle d|j'\rangle\langle j'|M_1\rangle\langle M_1|M_2\rangle\langle M_2|j'\rangle\langle j'|s\rangle\end{aligned} \tag{4.40}$$

leading to the same conclusions as with the Sagnac interferometer with three mirrors.

4.4 Beam-splitter matrices and Dirac's *bra ket* notation

A straightforward non-polarizing beam-splitter is a partial mirror. The 2×2 transfer matrix describing the action of a partial reflector, or beam splitter, is simply the identity matrix (Duarte 2003)

$$I = \begin{pmatrix} 1 & 0 \\ 0 & 1 \end{pmatrix} \tag{4.41}$$

If a photon polarized in the $|x\rangle$ state encounters a non-polarizing beam splitter, deployed at $\theta = \pi/4$, then there is a probability that the polarized photon will continue its straight path or that it will be reflected onto a path orthogonal to initial direction of propagation. No change in polarization is experienced.

Going back to the simple beam splitter with one input and two outputs, as described by equations (4.8) and (4.9),

$$|s\rangle_1 = |j\rangle\langle j|s\rangle \tag{4.42}$$

$$|s\rangle_2 = |j'\rangle\langle j'|s\rangle \tag{4.43}$$

Defining the vector states $|A\rangle$ and $|B\rangle$ as

$$|A\rangle = |j\rangle\langle j|s\rangle \tag{4.44}$$

$$|B\rangle = |j'\rangle\langle j'|s\rangle \tag{4.45}$$

then

$$|s\rangle_1 = |A\rangle \tag{4.46}$$

$$|s\rangle_2 = |B\rangle \tag{4.47}$$

which in vector form can be expressed as

$$\begin{pmatrix} |s\rangle_1 \\ |s\rangle_2 \end{pmatrix} = \begin{pmatrix} 1 & 0 \\ 0 & 1 \end{pmatrix} \begin{pmatrix} |A\rangle \\ |B\rangle \end{pmatrix} \tag{4.48}$$

with the identity matrix I representing the straightforward beam splitter.

4.4.1 The beam splitter and the Hadamard gate

Considering the quantum entanglement states given by (Pryce and Ward 1947, Duarte 2014)

$$|\psi\rangle_+ = 2^{-1/2}(|x\rangle_1|y\rangle_2 + |y\rangle_1|x\rangle_2) \tag{4.49}$$

$$|\psi\rangle_- = 2^{-1/2}(|x\rangle_1|y\rangle_2 - |y\rangle_1|x\rangle_2) \tag{4.50}$$

one can write (Dirac 1958; Duarte 2013)

$$|\alpha\rangle = |x\rangle_1|y\rangle_2 \tag{4.51}$$

$$|\beta\rangle = |y\rangle_1|x\rangle_2 \tag{4.52}$$

which leads directly to the quantum superposition equations describing beam splitter based interferometers such as the Mach–Zehnder interferometer (Duarte 2014, 2019)

$$|\psi\rangle_+ = 2^{-1/2}(|\alpha\rangle + |\beta\rangle) \tag{4.53}$$

$$|\psi\rangle_{-} = 2^{-1/2}(|\alpha\rangle - |\beta\rangle) \tag{4.54}$$

Notice that these equations have the exact form of equation (4.20) derived directly from the physics of the Mach–Zehnder interferometer relative to the output beam splitter with two orthogonal inputs.

Equations (4.53) and (4.54) can next be expressed in matrix form as

$$\begin{pmatrix} |\psi\rangle_{+} \\ |\psi\rangle_{-} \end{pmatrix} = 2^{-1/2}\begin{pmatrix} 1 & 1 \\ 1 & -1 \end{pmatrix}\begin{pmatrix} |\alpha\rangle \\ |\beta\rangle \end{pmatrix} \tag{4.55}$$

where

$$2^{-1/2}\begin{pmatrix} 1 & 1 \\ 1 & -1 \end{pmatrix} = H_G \tag{4.56}$$

is the *Hadamard gate*, as widely used in the literature.

4.5 Revisiting the single-beam splitter

The first two-beam interferometer, from a single source, was the Hanbury Brown and Twist (1956) interferometer, which utilized light emitted from the star Sirius as the light source (see figure 4.7). Starting from the Dirac–Feynman probability amplitude it can be shown that the interferometric superposition probability is given by equation (4.4) which for $N = 2$ becomes (Duarte 2014)

$$\langle d|s\rangle\langle d|s\rangle^{*} = \Psi(r_1)^2 + \Psi(r_2)^2 + 2\Psi(r_1)\Psi(r_2)\cos(\Omega_2 - \Omega_1) \tag{4.57}$$

and using $\Psi(r_1) = \Psi(r_2) = 1$, $\Omega_1 = kR_1$, $\Omega_2 = kR_2$

$$\langle d|s\rangle\langle d|s\rangle^{*} = 2 + \cos k(R_2 - R_1) \tag{4.58}$$

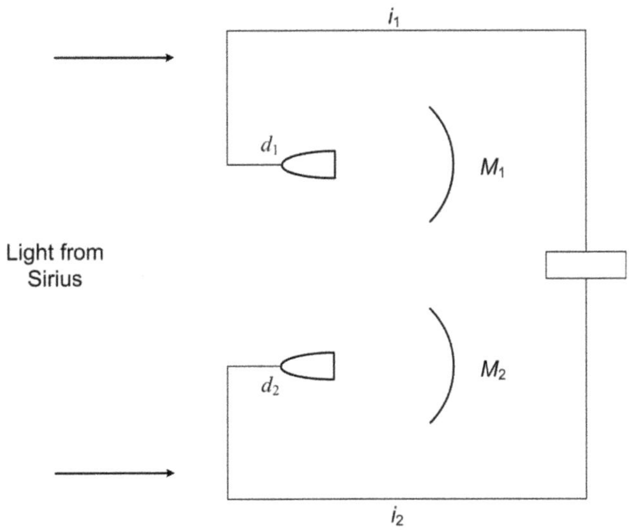

Figure 4.7. Schematics of the Hanbury Brown–Twist stellar interferometer.

as given by Feynman (Feynman *et al* 1965). Here, R_1 and R_2 are the distances from detector 1 and detector 2 to the stellar source.

It should be noted that an approximate equivalent optical diagram of figure 4.7 is that of a single-beam splitter with one input and two outputs, as illustrated in figure 4.2.

Assuming two identical detectors, this situation leads to a probability amplitude for $N = 2$ as given in

$$\langle d|s\rangle = \langle d|j'\rangle\langle j'|s\rangle + \langle d|j\rangle\langle j|s\rangle \tag{4.59}$$

which is the probability amplitude leading to the probability given in equation (4.57)

4.5.1 The HOM interferometer

The HOM interferometer (Hong *et al* 1987), illustrated in figure 4.8, can be outlined in an equivalent optical diagram, as illustrated in figure 4.9, assuming identical detectors $d_1 = d_2 = d$. It should also be noted that the incident photons are assumed to be *identical* and *indistinguishable* since interference between two distinguishable, or measurably distinct, photons does not occur (Dirac 1958). Photons emitted by two distinct lasers can be totally indistinguishable provided $\nu_1 = \nu_2$ and $\Delta\nu_1 = \Delta\nu_2$. In practice, indistinguishability might even extend to $\nu_1 \approx \nu_2$ provided the uncertainty in frequency is within $\Delta\nu_1 \approx \Delta\nu_2$.

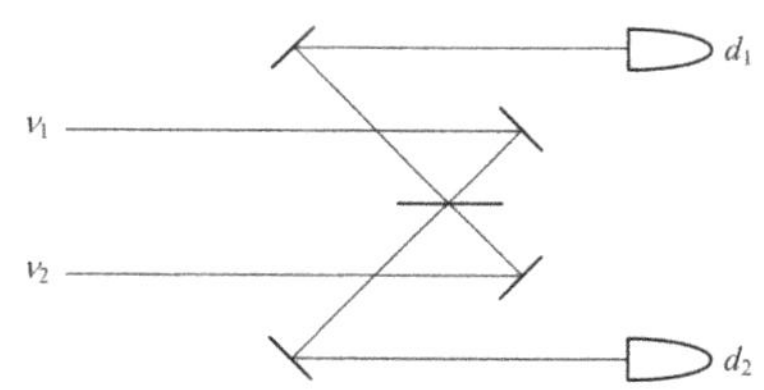

Figure 4.8. Simplified schematics of the HOM interferometer. The quanta pair ν_1 and ν_2 originate at the same source and $\nu_1 = \nu_2$.

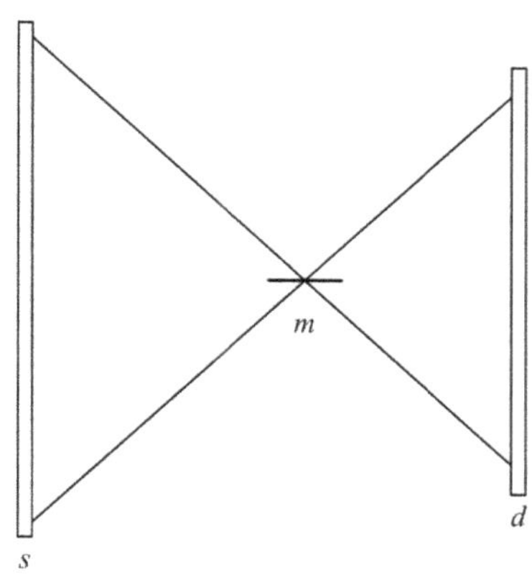

Figure 4.9. Equivalent optical diagram of the HOM interferometer assuming identical detectors, $d_1 = d_2$.

Previously the Dirac–Feynman interferometric principle (Dirac 1958, Feynman *et al* 1965)

$$\langle d|s\rangle = \sum_{j=1}^{N}\langle d|m\rangle\langle m|s\rangle \tag{4.60}$$

has been applied to derive the quantum entanglement probability amplitudes (Duarte 2013, 2014, 2019). Here, this principle is applied to the HOM configuration following the following convention: m is for the beam splitter in its reflection mode and m' is for the beam splitter in its transmission mode. Thus, following Duarte (2019)

$$\langle d|s\rangle = \langle d|m'\rangle\langle m'|s\rangle + \langle d|m\rangle\langle m|s\rangle \tag{4.61}$$

$$|s\rangle = |m'\rangle\langle m'|s\rangle + |m\rangle\langle m|s\rangle \tag{4.62}$$

which, using the Dirac identity $|\phi\rangle = |i\rangle\langle i|\phi\rangle$ can be abstracted as

$$|s\rangle = (|\gamma\rangle + |\delta\rangle) \tag{4.63}$$

Next, as done previously, using the Dirac identity

$$|X\rangle = |a\rangle_1|b\rangle_2|c\rangle_3\ldots|g\rangle_n \tag{4.64}$$

leads to

$$|\gamma\rangle = |x\rangle_1|y\rangle_2 \tag{4.65}$$

$$|\delta\rangle = |y\rangle_1|x\rangle_2 \tag{4.66}$$

so that

$$|\psi\rangle = (|x\rangle_1|y\rangle_2 + |y\rangle_1|x\rangle_2) \tag{4.67}$$

Taking into consideration its linear combination and normalizing lead again to the quantum entanglement states

$$|\psi\rangle_\pm = 2^{-1/2}(|x\rangle_1|y\rangle_2 \pm |y\rangle_1|x\rangle_2) \tag{4.68}$$

Going back to equation (4.62), normalizing and introducing its linear combination

$$|\psi\rangle_+ = 2^{-1/2}(|\gamma\rangle + |\delta\rangle) \tag{4.69}$$

$$|\psi\rangle_- = 2^{-1/2}(|\gamma\rangle - |\delta\rangle) \tag{4.70}$$

These equations can be expressed as

$$\begin{pmatrix}|\psi\rangle_+\\|\psi\rangle_-\end{pmatrix} = 2^{-1/2}\begin{pmatrix}1 & 1\\1 & -1\end{pmatrix}\begin{pmatrix}|\gamma\rangle\\|\delta\rangle\end{pmatrix} \tag{4.71}$$

Thus revealing, once again, the matrix for the Hadamard gate

$$H_G = 2^{-1/2}\begin{pmatrix} 1 & 1 \\ 1 & -1 \end{pmatrix}$$

which is widely associated as representing the quantum beam splitter and widely associated with the description of the HOM interferometer.

Problems

4.1 Expand equation (4.4) to arrive at the explicit interferometric equation for $N = 3$.
4.2 Expand equation (4.4) to arrive at the explicit interferometric equation for $N = 4$.
4.3 Show that equation (4.25) reduces to an equation of the form of equation (4.20).
4.4 Express the pair of equations (4.53) and (4.54) in matrix form to arrive at equation (4.55).
4.5 Starting from equation (4.1), derive the interferometric probability for the Hanbury Brown–Twist interferometer given in equation (4.57).

References

Born M 1926 Zur quantenmechanik der stoßvorgänge *Z. Phys.* **37** 863–27

Dirac P A M 1939 A new notation for quantum mechanics *Math. Proc. Camb. Phil. Soc.* **35** 416–8

Dirac P A M 1958 *The Principles of Quantum Mechanics* 4th edn (Oxford: Oxford University Press)

Duarte F J 1993 On a generalized interference equation and interferometric measurements *Opt. Commun.* **103** 8–14

Duarte F J 2003 *Tunable Laser Optics* (New York: Elsevier Academic)

Duarte F J 2013 The probability amplitude for entangled polarizations: an interferometric approach *J. Mod. Opt.* **60** 1585–7

Duarte F J 2014 *Quantum Optics for Engineers* (New York: CRC Press)

Duarte F J 2015 *Tunable Laser Optics* 2nd edn (New York: CRC Press)

Duarte F J 2019 *Fundamental of Quantum Entanglement* (Bristol: IOP Publishing)

Feynman R P, Leighton R B and Sands M 1965 *The Feynman Lectures on Physics* vol III (Reading, MA: Addison-Wesley)

Hanbury Brown R and Twist R Q 1956 A test of a new type of stellar interferometer on Sirius *Nature* **178** 1046–8

Hong C K, Ou Z Y and Mandel 1987 Measurements of subpicosecond time intervals between two photons by interference *Phys. Rev. Lett.* **59** 2044–6

Michelson A A 1927 *Studies in Optics* (Chicago, IL: The University of Chicago)

Pryce M L H and Ward J C 1947 Angular correlation effects with annihilation radiation *Nature* **160** 435

Steel W H 1976 *Interferometry* (Cambridge: Cambridge University Press)

IOP Publishing

Quantum Entanglement Engineering and Applications

F J Duarte and T S Taylor

Chapter 5

Vectors, matrices, and tensors for quantum entanglement

Vector products, matrices, and tensor elements applicable to quantum entanglement are reviewed in a direct and terse style. This review includes the vector and tensor products relevant to quantum entanglement probability amplitudes. Matrix algebra techniques directly applicable to quantum optics are described. Pauli matrices are utilized to illustrate the mathematics of eigenvalues and eigenvectors. Matrix unitary properties are also described.

5.1 Introduction

In this chapter aspects of vector algebra, and vector products, are described from a direct engineering style perspective. The description then integrates elements of matrix algebra and tensor algebra. The selective focus of this description includes vector, matrix, and tensor elements directly applicable to the mathematics of quantum entanglement.

This chapter is a review that brings together various subjects of interest previously treated in separate publications by Duarte (2014, 2019) and Duarte *et al* (2020). In this regard, this review offers a unified and improved treatment of the subject. For consistency the notation and style used in Duarte (2014, 2019) is preserved.

5.2 Vector basics

The sum of vectors in three dimensions (x, y, z), that is $\mathbf{u} + \mathbf{w}$, is defined as

$$\begin{pmatrix} u_1 \\ u_2 \\ u_3 \end{pmatrix} + \begin{pmatrix} w_1 \\ w_2 \\ w_3 \end{pmatrix} = \begin{pmatrix} u_1 + w_1 \\ u_2 + w_2 \\ u_3 + w_3 \end{pmatrix} \tag{5.1}$$

doi:10.1088/978-0-7503-3407-5ch5

and the subtraction $\mathbf{u} - \mathbf{w}$, is

$$\begin{pmatrix} u_1 \\ u_2 \\ u_3 \end{pmatrix} - \begin{pmatrix} w_1 \\ w_2 \\ w_3 \end{pmatrix} = \begin{pmatrix} u_1 - w_1 \\ u_2 - w_2 \\ u_3 - w_3 \end{pmatrix} \tag{5.2}$$

Multiplication of a vector $\mathbf{u}$ with a scalar number a, that is $a\mathbf{u}$, is given by

$$a\begin{pmatrix} u_1 \\ u_2 \\ u_3 \end{pmatrix} = \begin{pmatrix} au_1 \\ au_2 \\ au_3 \end{pmatrix} \tag{5.3}$$

The length of a vector $\mathbf{u}$, that is $|u|$, is calculated as

$$|\mathbf{u}|^2 = \begin{vmatrix} u_1 \\ u_2 \\ u_3 \end{vmatrix}^2 = (u_1^2 + u_2^2 + u_3^2) \tag{5.4}$$

5.3 Vector products

Vector products are widely used in quantum entanglement calculations. This section begins with the description of widely used vector products such as the dot product and the cross product and continues with the quantum specific *bra ket* product, the vector direct product, and the vector outer product.

5.3.1 Dot product

The dot product of two vectors $\mathbf{u} \cdot \mathbf{w}$ yields a scalar. The dot product is calculated as

$$\begin{pmatrix} u_1 \\ u_2 \\ u_3 \end{pmatrix} \cdot \begin{pmatrix} w_1 \\ w_2 \\ w_3 \end{pmatrix} = (u_1w_1 + u_2w_2 + u_3w_3) \tag{5.5}$$

For two vectors sustaining an angle θ between them, the dot product is given by

$$\mathbf{u} \cdot \mathbf{w} = |u||w| \cos\theta \tag{5.6}$$

5.3.2 Cross product

The cross product of two three-dimensional vectors leads to a new vector. The cross product $\mathbf{u} \times \mathbf{w}$ is defined as

$$\begin{pmatrix} u_1 \\ u_2 \\ u_3 \end{pmatrix} \times \begin{pmatrix} w_1 \\ w_2 \\ w_3 \end{pmatrix} = \begin{pmatrix} u_2w_3 - u_3w_2 \\ u_3w_1 - u_1w_3 \\ u_1w_2 - u_2w_1 \end{pmatrix} \tag{5.7}$$

5.3.3 The *ket bra* product

The product of a *ket* vector and a *bra* vector yields a matrix. This product is known as the *density matrix* ρ and is given by (Dirac 1958)

$$\rho = |u\rangle\ \langle u| \tag{5.8}$$

$$\rho = |u\rangle\ \langle u| = \begin{pmatrix} u_1 \\ u_2 \end{pmatrix} (u_1^* \quad u_2^*) = \begin{pmatrix} u_1 u_1^* & u_1 u_2^* \\ u_2 u_1^* & u_2 u_2^* \end{pmatrix} \tag{5.9}$$

where u_1^* and u_2^* are the complex conjugates of u_1 and u_2.

5.3.4 Vector direct product

The *vector direct product* is of crucial significance to the matrix description of quantum entanglement (Duarte 2019, Duarte *et al* 2020). Albeit the notation of the vector direct product also uses a *dot* in its nomenclature, it is different from the usual dot product.

The vector direct product is defined as (Ayres 1965)

$$|u\rangle|w\rangle = |u\rangle \cdot |w\rangle^T \tag{5.10}$$

$$|u\rangle \cdot |w\rangle^T = \begin{pmatrix} u_1 \\ u_2 \end{pmatrix} (w_1 \quad w_2) = \begin{pmatrix} u_1 w_1 & u_1 w_2 \\ u_2 w_1 & u_2 w_2 \end{pmatrix} \tag{5.11}$$

5.3.5 Vector outer product

An additional vector product of significance to quantum entanglement is the *vector outer product*. This product is sometimes associated in the literature with the symbol $\otimes$ but it is mathematically equivalent to the vector direct product (Ortega 1987)

$$|u\rangle|w\rangle^T = |u\rangle \otimes |w\rangle \tag{5.12}$$

$$|u\rangle \otimes |w\rangle = \begin{pmatrix} u_1 \\ u_2 \end{pmatrix} (w_1 \quad w_2) = \begin{pmatrix} u_1 w_1 & u_1 w_2 \\ u_2 w_1 & u_2 w_2 \end{pmatrix} \tag{5.13}$$

The symbol $\otimes$ is more widely recognized as being associated with the Kronecker product described in section 5.7.

5.4 Matrix algebra

The matrix mechanics described in this chapter applies to 2×2 matrices, 3×3 matrices and higher order matrices. However, for directness and simplicity the treatment focuses principally on 2×2 matrices.

Consider the 2×2 matrix A

$$A = \begin{pmatrix} a_{11} & a_{12} \\ a_{21} & a_{22} \end{pmatrix} \tag{5.14}$$

and the 2×2 matrix B

$$B = \begin{pmatrix} b_{11} & b_{12} \\ b_{21} & b_{22} \end{pmatrix} \tag{5.15}$$

The addition of these two matrices $A + B$ is defined as

$$A + B = \begin{pmatrix} a_{11} + b_{11} & a_{12} + b_{12} \\ a_{21} + b_{21} & a_{22} + b_{22} \end{pmatrix} \tag{5.16}$$

and the subtraction $A - B$ as

$$A - B = \begin{pmatrix} a_{11} - b_{11} & a_{12} - b_{12} \\ a_{21} - b_{21} & a_{22} - b_{22} \end{pmatrix} \tag{5.17}$$

Multiplication of the matrix A by (-1) yields the product

$$(-1)A = \begin{pmatrix} -a_{11} & -a_{12} \\ -a_{21} & -a_{22} \end{pmatrix} \tag{5.18}$$

Multiplication of the matrix A by the complex number i yields the product

$$iA = \begin{pmatrix} ia_{11} & ia_{12} \\ ia_{21} & ia_{22} \end{pmatrix} \tag{5.19}$$

The multiplication of these two matrices AB is defined as

$$AB = \begin{pmatrix} a_{11} & a_{12} \\ a_{21} & a_{22} \end{pmatrix}\begin{pmatrix} b_{11} & b_{12} \\ b_{21} & b_{22} \end{pmatrix} = \begin{pmatrix} a_{11}b_{11} + a_{12}b_{21} & a_{11}b_{12} + a_{12}b_{22} \\ a_{21}b_{11} + a_{22}b_{21} & a_{21}b_{12} + a_{22}b_{22} \end{pmatrix} \tag{5.20}$$

and the product BA is

$$BA = \begin{pmatrix} b_{11} & b_{12} \\ b_{21} & b_{22} \end{pmatrix}\begin{pmatrix} a_{11} & a_{12} \\ a_{21} & a_{22} \end{pmatrix} = \begin{pmatrix} b_{11}a_{11} + b_{12}a_{21} & b_{11}a_{12} + b_{12}a_{22} \\ b_{21}a_{11} + b_{22}a_{21} & b_{21}a_{12} + b_{22}a_{22} \end{pmatrix} \tag{5.21}$$

This leads directly to the conclusion

$$AB - BA \neq 0 \tag{5.22}$$

or

$$AB \neq BA \tag{5.23}$$

which is an iconic result in matrix algebra with an enormous impact in quantum physics.

For some special matrices the condition described by equations (5.22) and (5.23) might not hold as shall be illustrated next.

5.4.1 The identity matrix

In a 2×2 format the identity matrix I is defined as

$$I = \begin{pmatrix} 1 & 0 \\ 0 & 1 \end{pmatrix} \tag{5.24}$$

For 3×3 matrices

$$I = \begin{pmatrix} 1 & 0 & 0 \\ 0 & 1 & 0 \\ 0 & 0 & 1 \end{pmatrix} \tag{5.25}$$

and for 4×4 matrices

$$I = \begin{pmatrix} 1 & 0 & 0 & 0 \\ 0 & 1 & 0 & 0 \\ 0 & 0 & 1 & 0 \\ 0 & 0 & 0 & 1 \end{pmatrix} \tag{5.26}$$

Observe that for the unitary matrix product IA

$$IA = AI = A \tag{5.27}$$

so that, $IA - AI = 0$ which is an exception to the matrix product condition described in

$$AB - BA \neq 0.$$

5.4.2 The inverse matrix

The matrix C has an *inverse* matrix D if

$$C\ D = D\ C = I \tag{5.28}$$

defining the inverse matrix D as $D = C^{-1}$, equation (5.28) can also be expressed as

$$CC^{-1} = C^{-1}C = I \tag{5.29}$$

and

$$(C^{-1})^{-1} = C \tag{5.30}$$

5.4.3 The matrix determinant and trace

The *determinant* of the 2×2 matrix A, that is $|A|$, is given by

$$|A| = \begin{vmatrix} a_{11} & a_{12} \\ a_{21} & a_{22} \end{vmatrix} = a_{11}a_{22} - a_{21}a_{12} \tag{5.31}$$

and the determinant for a 3×3 matrix is calculated as

$$|A| = \begin{vmatrix} a_{11} & a_{12} & a_{13} \\ a_{21} & a_{22} & a_{23} \\ a_{31} & a_{32} & a_{33} \end{vmatrix} = a_{11} \begin{vmatrix} a_{22} & a_{23} \\ a_{32} & a_{33} \end{vmatrix} - a_{12} \begin{vmatrix} a_{21} & a_{23} \\ a_{31} & a_{33} \end{vmatrix} + a_{13} \begin{vmatrix} a_{21} & a_{22} \\ a_{31} & a_{32} \end{vmatrix} \tag{5.32}$$

The determinants of high order $N \times N$ matrices can be calculated applying the same mechanics.

An additional quantity used throughout quantum optics is the *trace* of a matrix designated as Tr. The trace of a matrix is defined as the sum of its diagonal elements

$$Tr(A) = \sum_j a_{jj} \tag{5.33}$$

More specifically, for a 3 × 3 matrix

$$Tr(A) = \sum_{j=1}^{N=3} a_{jj} = a_{11} + a_{22} + a_{33} \tag{5.34}$$

As an example consider the Hadamard matrix H and its inverse H^{-1}

$$H = \begin{pmatrix} 1 & 1 \\ 1 & -1 \end{pmatrix} \tag{5.35}$$

$$H^{-1} = \frac{1}{|H|}\begin{pmatrix} -1 & -1 \\ -1 & 1 \end{pmatrix} \tag{5.36}$$

$$H\ H^{-1} = \begin{pmatrix} 1 & 1 \\ 1 & -1 \end{pmatrix}(-2^{-1})\begin{pmatrix} -1 & -1 \\ -1 & 1 \end{pmatrix} = -2^{-1}\begin{pmatrix} -2 & 0 \\ 0 & -2 \end{pmatrix} = I \tag{5.37}$$

and the trace of this matrix is

$$Tr(H) = 1 + (-1) = 0 \tag{5.38}$$

5.4.4 Eigenvalues and eigenvectors

Useful concepts in matrix mechanics are those of eingenvalues and eigenvectors. For a matrix A the eigenvalues t and the eigenvector X

$$X = \begin{pmatrix} x \\ y \end{pmatrix} \tag{5.39}$$

are such that

$$AX = tX \tag{5.40}$$

or

$$\begin{pmatrix} a_{11} & a_{12} \\ a_{21} & a_{22} \end{pmatrix}\begin{pmatrix} x \\ y \end{pmatrix} = t\begin{pmatrix} x \\ y \end{pmatrix} \tag{5.41}$$

so that

$$a_{11}x + a_{12}y = tx \tag{5.42}$$

$$a_{21}x + a_{22}y = ty \tag{5.43}$$

which are equivalent to

$$(t - a_{11})x - a_{12}y = 0 \tag{5.44}$$

$$-a_{21}x + (t - a_{22})y = 0 \tag{5.45}$$

Writing these equations in matrix form and taking the determinant leads to

$$\begin{vmatrix} (t - a_{11}) & -a_{12} \\ -a_{21} & (t - a_{22}) \end{vmatrix} \tag{5.46}$$

$$(t - a_{11})(t - a_{22}) - a_{12}a_{21} = 0 \tag{5.47}$$

and the roots of this equation are the eigenvalues t. Using one of the eigenvalues in determinant (5.46) leads to a set of equations in x and y that yield one of the eigenvectors. The other eigenvalue leads to a second eigenvector. Examples are given in section 5.5.

5.5 The Pauli matrices

Here, Pauli matrices are considered from a purely mechanical perspective. The Pauli σ_x, σ_y, and σ_z matrices are given by (Dirac 1958)

$$\sigma_x = \begin{pmatrix} 0 & 1 \\ 1 & 0 \end{pmatrix} \tag{5.48}$$

$$\sigma_y = \begin{pmatrix} 0 & -i \\ i & 0 \end{pmatrix} \tag{5.49}$$

$$\sigma_z = \begin{pmatrix} 1 & 0 \\ 0 & -1 \end{pmatrix} \tag{5.50}$$

Some properties of these matrices are (Dirac 1958, Jordan 1986)

$$\sigma_x^2 = \sigma_y^2 = \sigma_z^2 = I \tag{5.51}$$

$$\sigma_x\sigma_y = i\sigma_z \tag{5.52}$$

$$\sigma_y\sigma_x = -i\sigma_z \tag{5.53}$$

$$\sigma_y\sigma_z = i\sigma_x \tag{5.54}$$

$$\sigma_z\sigma_y = -i\sigma_x \tag{5.55}$$

$$\sigma_z\sigma_x = i\sigma_y \tag{5.56}$$

$$\sigma_x\sigma_z = -i\sigma_y \tag{5.57}$$

$$\sigma_x\sigma_y\sigma_z = iI \tag{5.58}$$

The *conjugate* of σ_y is given by

$$\sigma_y^* = \begin{pmatrix} 0 & -i \\ +i & 0 \end{pmatrix}^* = \begin{pmatrix} 0 & +i \\ -i & 0 \end{pmatrix} \tag{5.59}$$

and the *transpose* (rows → column) of σ_y^* is $(\sigma_y^*)^T$

$$(\sigma_y^*)^T = \begin{pmatrix} 0 & +i \\ -i & 0 \end{pmatrix}^T = \begin{pmatrix} 0 & -i \\ +i & 0 \end{pmatrix} = \sigma_y \tag{5.60}$$

Equation (5.60) can be re expressed as

$$\sigma_y = \sigma_y^{\mathscr{H}} \tag{5.61}$$

where the symbol $\mathscr{H}$ here represents the combined *conjugate transpose* operation also known as the *Hermitian property*. All three Pauli matrices are Hermitian.

Furthermore, it should be noticed that from equations (5.35), (5.48), and (5.50) one can write (Duarte *et al* 2020)

$$H = \sigma_x + \sigma_z \tag{5.62}$$

5.5.1 Eigenvalues of the Pauli matrices

Consider σ_x and follow the methodology of section 5.4.4. This leads to the system

$$\begin{pmatrix} 0 & 1 \\ 1 & 0 \end{pmatrix}\begin{pmatrix} x \\ y \end{pmatrix} = t\begin{pmatrix} x \\ y \end{pmatrix} \tag{5.63}$$

$$0x + 1y = tx \tag{5.64}$$

$$1x + 0y = ty \tag{5.65}$$

$$\begin{vmatrix} t & -1 \\ -1 & t \end{vmatrix} \tag{5.66}$$

$$t^2 - 1 = 0 \tag{5.67}$$

$$t = \pm 1 \tag{5.68}$$

Substituting for $t = +1$, into the matrix (5.66),

$$\begin{pmatrix} +1 & -1 \\ -1 & +1 \end{pmatrix} \tag{5.69}$$

leading to

$$x - y = 0 \tag{5.70}$$

$$-x + y = 0 \tag{5.71}$$

so that an eigenvector is

$$X_1 = \begin{pmatrix} x \\ y \end{pmatrix} = \begin{pmatrix} 1 \\ 1 \end{pmatrix} \tag{5.72}$$

Similarly for $t = -1$, the second eigenvector can be

$$X_2 = \begin{pmatrix} 1 \\ -1 \end{pmatrix} \tag{5.73}$$

Now, consider σ_y:

$$\begin{pmatrix} 0 & -i \\ i & 0 \end{pmatrix}\begin{pmatrix} x \\ y \end{pmatrix} = t\begin{pmatrix} x \\ y \end{pmatrix} \tag{5.74}$$

$$0x - iy = tx \tag{5.75}$$

$$ix + 0y = ty \tag{5.76}$$

$$\begin{vmatrix} t & +i \\ -i & t \end{vmatrix} \tag{5.77}$$

$$t^2 - 1 = 0 \tag{5.78}$$

$$t = \pm 1 \tag{5.79}$$

A set of corresponding eigenvectors, for $t = \pm 1$, is

$$X_1 = \begin{pmatrix} -i \\ 1 \end{pmatrix} \tag{5.80}$$

$$X_2 = \begin{pmatrix} 1 \\ -i \end{pmatrix} \tag{5.81}$$

Next, consider the matrix $(\sigma_x + \sigma_y)$:

$$\begin{pmatrix} 0 & 1-i \\ 1+i & 0 \end{pmatrix}\begin{pmatrix} x \\ y \end{pmatrix} = t\begin{pmatrix} x \\ y \end{pmatrix} \tag{5.82}$$

$$0x + (1-i)y = tx \tag{5.83}$$

$$(1+i)x + 0y = ty \tag{5.84}$$

$$\begin{vmatrix} -t & (1-i) \\ (1+i) & -t \end{vmatrix} \tag{5.85}$$

$$t^2 - 2 = 0 \tag{5.86}$$

$$t = \pm\sqrt{2} \tag{5.87}$$

A set of corresponding eigenvectors, for $t = \pm\sqrt{2}$, is

$$X_1 = \begin{pmatrix} 2^{-1/2}(1 - i) \\ 1 \end{pmatrix} \tag{5.88}$$

$$X_2 = \begin{pmatrix} 1 \\ -2^{-1/2}(1 + i) \end{pmatrix} \tag{5.89}$$

An important result from this exercise is

$$t_{\sigma_x} + t_{\sigma_y} \neq t_{\sigma_x + \sigma_y} \tag{5.90}$$

In words this means that the sum of the eigenvalue of σ_x plus the eigenvalue of σ_y is different to the eigenvalue of the sum of the matrices $(\sigma_x + \sigma_y)$. This is the principal argument used by Bell (1964) to question the relevance of von Neumann's argument (von Neumann 1932).

5.6 Unitary matrices

A complex matrix G, obeying the property

$$G\ G^* = G^* G = I \tag{5.91}$$

is considered to be a *unitary* matrix (Lipschutz 1968). For example if

$$G = \begin{pmatrix} i & 0 \\ 0 & i \end{pmatrix} \tag{5.92}$$

then

$$G\ G^* = \begin{pmatrix} i & 0 \\ 0 & i \end{pmatrix}\begin{pmatrix} -i & 0 \\ 0 & -i \end{pmatrix} = I \tag{5.93}$$

Here it should be noted that some authors define a unitary matrix as a matrix J exhibiting the more stringent property (Tropper 1968)

$$(J^*)^T J = J(J^*)^T = I \tag{5.94}$$

In this regard, if

$$J = \begin{pmatrix} +i & 0 \\ 0 & -i \end{pmatrix} \tag{5.95}$$

$$(J^*)^T = \begin{pmatrix} -i & 0 \\ 0 & +i \end{pmatrix} \tag{5.96}$$

and the condition imposed in equation (5.94) is met, namely

$$\begin{pmatrix} -i & 0 \\ 0 & +i \end{pmatrix}\begin{pmatrix} +i & 0 \\ 0 & -i \end{pmatrix} = \begin{pmatrix} +i & 0 \\ 0 & -i \end{pmatrix}\begin{pmatrix} -i & 0 \\ 0 & +i \end{pmatrix} = \begin{pmatrix} 1 & 0 \\ 0 & 1 \end{pmatrix} \tag{5.97}$$

5.7 The tensor product

This section is based on the brief introduction given by Duarte (2019) and it is specifically tailored to deal with tensor elements applicable to the mathematics of the probability amplitude of quantum entanglement.

The *tensor product*, also known as the Kronecker product $\mathbf{U} \otimes \mathbf{W}$, is a vector-matrix multiplication in which each element of the product vector or matrix is comprised of each of the elements of the $\mathbf{U}$ vector or matrix, u_{mn}, multiplying the whole $\mathbf{W}$ vector or matrix.

Thus, the first element of the tensor product is $u_{11}\mathbf{W}$, and the last element is $u_{mn}\,\mathbf{W}$ (Zehfuss 1858). This means, for example, that the Kronecker product of two 2×2 matrices yields a 4×4 matrix. For two-dimensional vectors this product can be expressed as

$$|u\rangle \otimes |w\rangle = \begin{pmatrix} u_1 \\ u_2 \end{pmatrix} \otimes \begin{pmatrix} w_1 \\ w_2 \end{pmatrix} = \begin{pmatrix} u_1 w_1 \\ u_1 w_2 \\ u_2 w_1 \\ u_2 w_2 \end{pmatrix} \tag{5.98}$$

For the vectors

$$|1\rangle = \begin{pmatrix} 1 \\ 0 \end{pmatrix} \tag{5.99}$$

$$|0\rangle = \begin{pmatrix} 0 \\ 1 \end{pmatrix} \tag{5.100}$$

the following Kronecker products follow:

$$|1\rangle \otimes |1\rangle = \begin{pmatrix} 1 \\ 0 \\ 0 \\ 0 \end{pmatrix} \tag{5.101}$$

$$|1\rangle \otimes |0\rangle = \begin{pmatrix} 0 \\ 1 \\ 0 \\ 0 \end{pmatrix} \tag{5.102}$$

$$|0\rangle \otimes |1\rangle = \begin{pmatrix} 0 \\ 0 \\ 1 \\ 0 \end{pmatrix} \tag{5.103}$$

$$|0\rangle \otimes |0\rangle = \begin{pmatrix} 0 \\ 0 \\ 0 \\ 1 \end{pmatrix} \tag{5.104}$$

For 2×2 matrices the Kronecker product $\mathbf{U} \otimes \mathbf{W}$ is defined as

$$\begin{pmatrix} u_{11} & u_{12} \\ u_{21} & u_{22} \end{pmatrix} \otimes \begin{pmatrix} w_{11} & w_{12} \\ w_{21} & w_{22} \end{pmatrix} = \begin{pmatrix} u_{11}w_{11} & u_{11}w_{12} & u_{12}w_{11} & u_{12}w_{12} \\ u_{11}w_{21} & u_{11}w_{22} & u_{12}w_{21} & u_{12}w_{22} \\ u_{21}w_{11} & u_{21}w_{12} & u_{22}w_{12} & u_{22}w_{12} \\ u_{21}w_{21} & u_{21}w_{22} & u_{22}w_{21} & u_{22}w_{22} \end{pmatrix} \tag{5.105}$$

Using the Kronecker product configuration on the Pauli matrices yield the following 4×4 matrices

$$\sigma_x \otimes \sigma_y = \begin{pmatrix} 0 & 1 \\ 1 & 0 \end{pmatrix} \otimes \begin{pmatrix} 0 & -i \\ i & 0 \end{pmatrix} = \begin{pmatrix} 0 & 0 & 0 & -i \\ 0 & 0 & i & 0 \\ 0 & -i & 0 & 0 \\ i & 0 & 0 & 0 \end{pmatrix} \tag{5.106}$$

$$\sigma_x \otimes \sigma_z = \begin{pmatrix} 0 & 1 \\ 1 & 0 \end{pmatrix} \otimes \begin{pmatrix} 1 & 0 \\ 0 & -1 \end{pmatrix} = \begin{pmatrix} 0 & 0 & 1 & 0 \\ 0 & 0 & 0 & -1 \\ 1 & 0 & 0 & 0 \\ 0 & -1 & 0 & 0 \end{pmatrix} \tag{5.107}$$

$$\sigma_y \otimes \sigma_x = \begin{pmatrix} 0 & -i \\ i & 0 \end{pmatrix} \otimes \begin{pmatrix} 0 & 1 \\ 1 & 0 \end{pmatrix} = \begin{pmatrix} 0 & 0 & 0 & -i \\ 0 & 0 & -i & 0 \\ 0 & i & 0 & 0 \\ i & 0 & 0 & 0 \end{pmatrix} \tag{5.108}$$

$$\sigma_y \otimes \sigma_z = \begin{pmatrix} 0 & -i \\ i & 0 \end{pmatrix} \otimes \begin{pmatrix} 1 & 0 \\ 0 & -1 \end{pmatrix} = \begin{pmatrix} 0 & 0 & -i & 0 \\ 0 & 0 & 0 & i \\ i & 0 & 0 & 0 \\ 0 & -i & 0 & 0 \end{pmatrix} \tag{5.109}$$

$$\sigma_z \otimes \sigma_x = \begin{pmatrix} 1 & 0 \\ 0 & -1 \end{pmatrix} \otimes \begin{pmatrix} 0 & 1 \\ 1 & 0 \end{pmatrix} = \begin{pmatrix} 0 & 1 & 0 & 0 \\ 1 & 0 & 0 & 0 \\ 0 & 0 & 0 & -1 \\ 0 & 0 & -1 & 0 \end{pmatrix} \tag{5.110}$$

$$\sigma_z \otimes \sigma_y = \begin{pmatrix} 1 & 0 \\ 0 & -1 \end{pmatrix} \otimes \begin{pmatrix} 0 & -i \\ i & 0 \end{pmatrix} = \begin{pmatrix} 0 & -i & 0 & 0 \\ i & 0 & 0 & 0 \\ 0 & 0 & 0 & i \\ 0 & 0 & -i & 0 \end{pmatrix} \tag{5.111}$$

Problems

5.1 Prove that $\sigma_x^2 = \sigma_y^2 = \sigma_z^2 = I$.
5.2 Prove that $\sigma_y\sigma_x = -i\sigma_z$.
5.3 Prove that $\sigma_x\sigma_y\sigma_z = iI$.
5.4 Prove that $H = \sigma_x + \sigma_z$.

5.5 Show that a set of eigenvectors for $(\sigma_x + \sigma_y)$ is given in equations (5.88) and (5.89).
5.6 Verify that $|1\rangle \otimes |1\rangle$ is given by equation (5.101).
5.7 Verify that $\sigma_x \otimes \sigma_y$ is given by equation (5.106).
5.8 Verify that $\sigma_z \otimes \sigma_y$ is given by equation (5.111).

References

Ayres F 1965 *Modern Algebra* (New York: McGraw-Hill)
Bell J S 1964 On the Einstein–Podolsky–Rosen paradox *Physics (N. Y.)* **1** 195–200
Dirac P A M 1958 *The Principles of Quantum Mechanics* 4th edn (Oxford: Oxford University Press)
Duarte F J 2014 *Quantum Optics for Engineers* (New York: CRC Press)
Duarte F J 2019 *Fundamentals of Quantum Entanglement* (Bristol: IOP Publishing)
Duarte F J, Taylor T S and Slaten J C 2020 On the probability amplitude of quantum entanglement and the Pauli matrices *Opt. Quantum Electron.* **52** 106
Jordan T E 1986 *Quantum Mechanics in Simple Matrix Form* (New York: Wiley)
Lipschutz S 1968 *Linear Algebra* (New York: McGraw Hill)
Ortega J M 1987 *Matrix Theory* (New York: Plenum)
Tropper A M 1968 *Linear Algebra* (Nelson: London)
von Neumann J 1932 *Mathematische Grundlagen der Quanten-Mechanik* (Berlin: Springer)
Zehfuss G 1858 Über eine gewisse determinante *Z. Math. Phys.* **3** 298–301

IOP Publishing

Quantum Entanglement Engineering and Applications

F J Duarte and T S Taylor

Chapter 6

Five avenues to the probability amplitude of quantum entanglement

The five paths to quantum entanglement are discussed quantitatively. Starting with Ward's heuristic derivation, the chapter continues to describe Feynman's two-state approach, the N-slit interferometric approach, the Pauli matrix approach, and the Hadarmard matrix approach. The chapter concludes with a discussion on the fundamentality of quantum entanglement versus the fundamentality of quantum interference at the foundations of quantum mechanics.

6.1 Introduction

In this chapter five derivational paths leading to the superposition probability amplitude of quantum entanglement are discussed quantitatively. First, Ward's heuristic approach is described as exhibited in his 1949 doctoral thesis; secondly, attention is focused on the fairly technical Feynman's two-state approach. Then the chapter continues to describe the N-slit interferometric approach, the Pauli matrix approach, and the Hadarmard matrix approach. In this regard, this chapter is based on the ideas and exposition already presented in *Fundamentals of Quantum Entanglement* (Duarte 2019) but does so in a unified and expanded manner while at the same time providing elucidating new perspectives.

The chapter concludes with a discussion on the fundamental significance of quantum entanglement versus the fundamental significance of quantum interference at the foundations of quantum mechanics.

6.2 Ward's heuristic derivation

Although the probability amplitude for quantum entanglement was not explicitly included in the landmark paper by Pryce and Ward (1947), it was included in its superposition probabilistic version in the final scattering cross section equation

doi:10.1088/978-0-7503-3407-5ch6

$$\frac{r_0^4 d\Omega_1 d\Omega_2}{16}\left[\frac{((1-\cos\theta_1)^3+2)((1-\cos\theta_2)^3+2)}{(2-\cos\theta_1)^3(2-\cos\theta_2)^3} - \frac{(\sin^2\theta_1 \sin^2\theta_2)\cos 2(\varphi_1-\varphi_2)}{(2-\cos\theta_1)^2(2-\cos\theta_2)^2}\right] \tag{6.1}$$

which for the useful experimental setting of $\theta_1 = \theta_2 = \theta/2$ can be expressed as

$$\sigma(\varphi_1, \varphi_2) = K\big(A + B(-\cos 2(\varphi_1 - \varphi_2))\big) \tag{6.2}$$

where K is a constant and A and B are real numbers. In this equation the term $(-\cos 2(\varphi_1 - \varphi_2))$ is the superposition quantum probability

$$P(\varphi_1, \varphi_2) = -\cos 2(\varphi_1 - \varphi_2) \tag{6.3}$$

This quantum probability is obtained directly from the *correct* probability amplitude of two quanta with entangled polarizations, as shown in chapter 10 (Duarte 2019). With this insight in mind attention is now given to Ward's brief heuristic derivation of the probability amplitude for quantum entanglement.

Ward (1949) confronted the problem from a conservation of momentum perspective.

Singlet states, as suggested from Wheeler's description (Wheeler 1946), were also central to the argument. In his writings, Ward begins with: 'Wheeler proceeded to calculate… But through the neglect of interference terms he derived an incorrect… far too small value for the angular correlations…' (Ward 1949).

He continues: '*it is essential to derive correctly the state vector which properly describes the state of the two quanta*' (Ward 1949). By the correct 'state vector' Ward means the correct equation for the superposition probability amplitude for quantum entanglement.

In this task Ward next listed the polarization state alternatives for the $|x\rangle$ and $|y\rangle$ polarizations states involved in a configuration of two counter propagating quanta:

$$|x, x\rangle, |x, y\rangle, |y, x\rangle, |y, y\rangle \tag{6.4}$$

He then considered the momenta options depicted in figure 6.1 (Ward 1949)

$$|+k, -k\rangle, |-k, +k\rangle \tag{6.5}$$

According to Ward there are singlet and triple state options for the combined states.

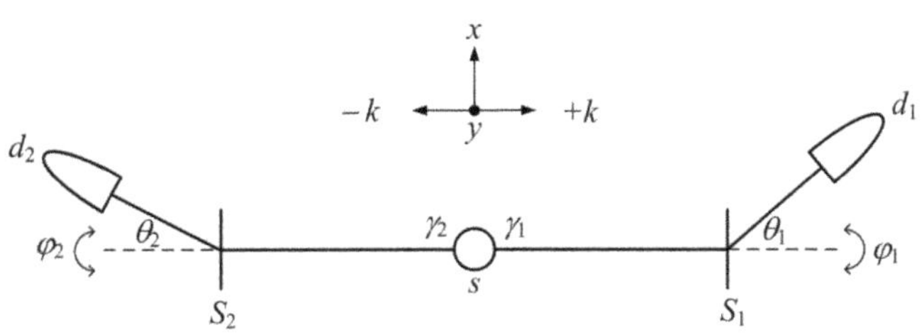

Figure 6.1. Version of the original Pryce–Ward experimental outline. In this diagram, s is the e^+e^- annihilation source of the γ_1 and γ_2 quanta. These quanta are emitted with $+k$ and $-k$ momenta (Duarte 2019).

He argues that triplet state configuration includes states symmetrical in polarization and momenta. Ward dismisses triplet states since the only admissible option, in the overall configuration, includes quanta with parallel polarization.

Thus, Ward arrives to the singlet state option which is anti-symmetrical both in polarization and momenta (Ward 1949)

$$(|x, y\rangle - |y, x\rangle)(|+k, -k\rangle - |-k, +k\rangle) \tag{6.6}$$

Therefore, the probability amplitude describing the interaction of the combined polarization states is

$$(|x, y\rangle - |y, x\rangle) \tag{6.7}$$

That's it! This superposition probability amplitude, or state, has all the physics of quantum entanglement in it. The rest is just the clever use of Dirac's notation and some mathematical formalism.

Attaching the numeral 1 to the photon associated with the first coordinate and the numeral 2 to the photon associated with the second coordinate then the states expressed in (6.4) can be rewritten as

$$|x_1, x_2\rangle, |x_1, y_2\rangle, |y_1, x_2\rangle, |y_1, y_2\rangle \tag{6.8}$$

and expression (6.6) can be written as

$$(|x_1, y_2\rangle - |y_1, x_2\rangle) \tag{6.9}$$

Applying the Dirac identity $|x, y\rangle = |x\rangle|y\rangle$, this expression can be recast more formally as the probability amplitude (Duarte 2014)

$$|\psi\rangle = (|x\rangle_1 |y\rangle_2 - |y\rangle_1 |x\rangle_2) \tag{6.10}$$

which, following straightforward normalization, becomes

$$|\psi\rangle = 2^{-1/2}(|x\rangle_1 |y\rangle_2 - |y\rangle_1 |x\rangle_2) \tag{6.11}$$

This is the iconic equation for the probability amplitude of entangled polarizations of quanta propagating in opposite directions. John Clive Ward was merely 23 years old when he did this extraordinary piece of clever physics.

6.2.1 The quantum entanglement probability

Using the probability amplitude given in equation (6.11), and following an analysis of the polarization alternatives (Duarte 2014, 2019), it can be shown that

$$|\psi\rangle = 2^{-1/2}(\sin\varphi_1 \cos\varphi_2 - \cos\varphi_1 \sin\varphi_2) \tag{6.12}$$

$$|\psi\rangle = 2^{-1/2}\sin(\varphi_1 - \varphi_2) \tag{6.13}$$

and if the polarization analyzers are rotated by $\pi/2$

$$|\psi\rangle = 2^{-1/2}\cos(\varphi_1 - \varphi_2) \tag{6.14}$$

Denoting transmission through a polarization analyzer by + and absorption by − (Mandel and Wolf 1995) then the corresponding probabilities are

$$p(+\varphi_1, +\varphi_2) = p(-\varphi_1, -\varphi_2) = 2^{-1}\sin^2(\varphi_1 - \varphi_2) \tag{6.15}$$

$$p(+\varphi_1, -\varphi_2) = p(-\varphi_1, +\varphi_2) = 2^{-1}\cos^2(\varphi_1 - \varphi_2) \tag{6.16}$$

The overall probability $P(\varphi_1, \varphi_2)$ is the sum of the individual probability alternatives (Mandel and Wolf 1995)

$$P(\varphi_1, \varphi_2) = p(+\varphi_1, +\varphi_2) + p(-\varphi_1, -\varphi_2) - p(+\varphi_1, -\varphi_2) - p(-\varphi_1, +\varphi_2) \tag{6.17}$$

$$P(\varphi_1, \varphi_2) = \sin^2(\varphi_1 - \varphi_2) - \cos^2(\varphi_1 - \varphi_2) \tag{6.18}$$

$$P(\varphi_1, \varphi_2) = -\cos 2(\varphi_1 - \varphi_2) \tag{6.19}$$

as previously given in equation (6.3) (Pryce and Ward 1947).

6.3 Quantum entanglement from Feynman's two-state approach

While discussing the dynamics of a two-state system via

$$i\hbar\frac{dC_i}{dt} = \sum H_{ij}C_j \tag{6.20}$$

where H_{ij} is the Hamiltonian, Feynman (Feynman *et al* 1965) discusses the generalized Dirac principle

$$\langle\psi|\phi\rangle = \sum_{i=1}\langle\psi|i\rangle\langle i|\phi\rangle \tag{6.21}$$

Setting $\phi = \psi = II$, for $i = 1, 2$

$$\langle II|II\rangle = \langle II|1\rangle\langle 1|II\rangle + \langle II|2\rangle\langle 2|II\rangle \tag{6.22}$$

$$\langle II|II\rangle = \langle II|1\rangle\langle II|1\rangle^* + \langle II|2\rangle\langle II|2\rangle^* \tag{6.23}$$

$$\langle II|II\rangle = C_I C_I^* + C_{II} C_{II}^* \tag{6.24}$$

$$\langle II|II\rangle = |C_I|^2 + |C_{II}|^2 \tag{6.25}$$

and since $\langle II|II\rangle = 1$

$$|C_I|^2 + |C_{II}|^2 = 1 \tag{6.26}$$

Expressing C_I and C_{II} in terms of C_1 and C_2, lead to the following normalized expressions

$$C_{II} = 2^{-1/2}(C_1 + C_2) \tag{6.27}$$

$$C_I = 2^{-1/2}(C_1 - C_2) \tag{6.28}$$

so that

$$C_1^2 + C_2^2 = 1 \tag{6.29}$$

Setting $C_1 = \langle 1|\phi\rangle$ and $C_2 = \langle 2|\phi\rangle$, equations (6.27) and (6.28) can be written as

$$\langle II|\phi\rangle = 2^{-1/2}(\langle 1|\phi\rangle + \langle 2|\phi\rangle) \tag{6.30}$$

$$\langle I|\phi\rangle = 2^{-1/2}(\langle 1|\phi\rangle - \langle 2|\phi\rangle) \tag{6.31}$$

Abstraction of $|\phi\rangle$ in equation (6.31) leads to

$$\langle I| = 2^{-1/2}(\langle 1| - \langle 2|) \tag{6.32}$$

which is the same as (Feynman *et al* 1965)

$$|I\rangle = 2^{-1/2}(|1\rangle - |2\rangle) \tag{6.33}$$

Using the same methodology it can also be shown that

$$|II\rangle = 2^{-1/2}(|1\rangle + |2\rangle) \tag{6.34}$$

This is where Feynman left it in 1965 in his discussion of the dynamics of a two-state system (Feynman *et al* 1965).

As pointed out by Duarte (2014), equation (6.33) can be more generally written as

$$|\psi\rangle = 2^{-1/2}(|B\rangle - |A\rangle) \tag{6.35}$$

For two quanta, $|B\rangle$ and $|A\rangle$ can be combined states. Using the Dirac identity $|X\rangle = |a\rangle_1\,|b\rangle_2\,|c\rangle_3 \ldots |g\rangle_n$ (Dirac 1958), $|B\rangle$ and $|A\rangle$ can be expressed as

$$|B\rangle = |x\rangle_1\,|y\rangle_2 \tag{6.36}$$

$$|A\rangle = |y\rangle_1\,|x\rangle_2 \tag{6.37}$$

Substitution of the identities (6.36) and (6.37) into equation (6.35) leads directly to the original probability amplitude for quantum entanglement (Duarte 2014)

$$|\psi\rangle = 2^{-1/2}(|x\rangle_1\,|y\rangle_2 - |y\rangle_1\,|x\rangle_2) \tag{6.38}$$

Further on in his lectures, Feynman discusses the hyperfine splitting of hydrogen and, without explanation, he allows whole states to be described by notations like $|+\,+\rangle$, $|+\,-\rangle$, $|-\,+\rangle$, and $|-\,-\rangle$, which allows the expression of probability amplitudes like (Feynman *et al* 1965)

$$|IV\rangle = 2^{-1/2}(|+\,-\rangle - |-\,+\rangle) \tag{6.39}$$

Later in his discussion of quantum entanglement via the annihilation of the positronium, again without explanation, and only using the cryptic sentence 'theoretical physicists have shown' he introduces expressions like

$$|F\rangle = (|R_1R_2\rangle - |L_1L_2\rangle) \tag{6.40}$$

The keen observer would immediately realize that equations (6.39) and (6.40) follow the notation style introduced by Ward (1949).

One important footnote is due here: equations of the form of Feynman's equation (6.39) appeared in the literature of quantum entanglement in the early 1990s with neither reference to the original source nor to Feynman.

6.4 Quantum entanglement from *N*-slit interference

The optical version of a Pryce and Ward type experimental diagram for quantum entanglement is shown in figure 6.1. Here, s is the quanta source of ν_1 and ν_2 emitted in the $+\ z$- and $-\ z$-directions, respectively. Photon propagation occurs along the optical axis, in the $+\ z$- and $-\ z$-directions. The photons pass through polarization analyzers p_1 and p_2 toward detectors d_1 and d_2, respectively.

For the emission of *indistinguishable* quanta, $\nu_1 = \nu_2 = \nu$, and matched ideal detectors $d_1 = d_2 = d$, the photon propagation is described by Dirac interferometric principle

$$\langle d|s\rangle = \sum_{j=1}^{N} \langle d|p_j\rangle\langle p_j|s\rangle \tag{6.41}$$

$$\langle d|s\rangle = \langle d|p_2\rangle\langle p_2|s\rangle + \langle d|p_1\rangle\langle p_1|s\rangle \tag{6.42}$$

Abstracting $\langle d$ from equation (6.41), leads to

$$|s\rangle = |p_1\rangle\langle p_2|s\rangle + |p_1\rangle\langle p_1|s\rangle \tag{6.43}$$

Using the Dirac identity (see chapter 2)

$$|\psi\rangle = |j\rangle\langle j|\psi\rangle \tag{6.44}$$

the two terms on the RHS of equation (6.43) can be written as

$$|D\rangle_2 = |p_2\rangle\langle p_2|s\rangle \tag{6.45}$$

$$|D\rangle_1 = |p_1\rangle\langle p_1|s\rangle \tag{6.46}$$

Substituting identities (6.45) and (6.46) into equation (6.43) yields

$$|s\rangle = (|D\rangle_2 + |D\rangle_1) \tag{6.47}$$

Equation (6.47) is the probability amplitude $|s\rangle$ which is the linear combination of two other probability amplitudes, or states, $|D\rangle_1$ and $|D\rangle_2$. This probability amplitude applies naturally to the description of an interferometric situation involving a single photon, or an ensemble of indistinguishable photons. It also applies to the quantum entanglement framework being developed here.

The crucial differentiation between interference and quantum entanglement occurs at this very stage that brings in the Dirac identity for indistinguishable quanta (see chapters 2 and 3)

$$|X\rangle = |a\rangle_1\,|b\rangle_2\,|c\rangle_3\ldots|g\rangle_n \tag{6.48}$$

This very step diverges from a purely interferometric situation given that the possibility of quantum 1 and quantum 2, is allowed with each photon in different alternative states. That is, state $|x\rangle$ and state $|y\rangle$ (Duarte 2019). That's all the physics.

To distinguish the notation of quantum entanglement from a pure interferometric situation $|D\rangle_1$ and $|D\rangle_2$ are renamed $|C\rangle_1$ and $|C\rangle_2$ so that equation (6.47) becomes

$$|s\rangle = (|C\rangle_2 + |C\rangle_1) \tag{6.49}$$

where

$$|C\rangle_2 = |x\rangle_1\,|y\rangle_2 \tag{6.50}$$

and

$$|C\rangle_1 = |y\rangle_1\,|x\rangle_2 \tag{6.51}$$

Using the Dirac identities (6.50) and (6.51) in the probability amplitude (6.49) yields (Duarte 2013a, 2013b, 2014)

$$|s\rangle = (|x\rangle_1\,|y\rangle_2 + |y\rangle_1\,|x\rangle_2) \tag{6.52}$$

Designating $|s\rangle$ as $|\psi\rangle$, and normalizing, equation (6.52) can be expressed as

$$|\psi\rangle_+ = 2^{-1/2}(|x\rangle_1\,|y\rangle_2 + |y\rangle_1\,|x\rangle_2) \tag{6.53}$$

and its linear combination is

$$|\psi\rangle_- = 2^{-1/2}(|x\rangle_1\,|y\rangle_2 - |y\rangle_1\,|x\rangle_2) \tag{6.54}$$

which is the iconic probability amplitude for quantum entanglement.

For notational convenience, the linear combinations (6.53) and (6.54) can be described in a 2 × 2 arrangement

$$\begin{matrix} +|C\rangle_1 & +|C\rangle_2 \\ +|C\rangle_1 & -|C\rangle_2 \end{matrix} \tag{6.55}$$

6.4.1 Additional pair of quantum entanglement states

The case for pairs with identical states of polarization has been treated by Duarte (2019).

In equation (6.49), if the two quanta pairs have identical states of polarization, that is, $|x\rangle_1\,|x\rangle_2$ and $|y\rangle_1\,|y\rangle_2$, then the identities (6.50) and (6.51) become (see chapter 3)

$$|C\rangle_2 = |x\rangle_1\,|x\rangle_2 \tag{6.56}$$

and

$$|C\rangle_1 = |y\rangle_1\,|y\rangle_2 \tag{6.57}$$

so that

$$|\psi\rangle^{+} = 2^{-1/2}(|x\rangle_1\,|x\rangle_2 + |y\rangle_1\,|y\rangle_2) \tag{6.58}$$

and

$$|\psi\rangle^{-} = 2^{-1/2}(|x\rangle_1\,|x\rangle_2 - |y\rangle_1\,|y\rangle_2) \tag{6.59}$$

These types of probability amplitudes, $|\psi\rangle^{+}$ and $|\psi\rangle^{-}$, are of interest in quantum computing. In summary, the whole family of probability amplitudes relevant to two quanta is represented by $|\psi\rangle_{+}$, $|\psi\rangle_{-}$, $|\psi\rangle^{+}$, and $|\psi\rangle^{-}$

$$\begin{aligned}
|\psi\rangle_{+} &= 2^{-1/2}(|x\rangle_1\,|y\rangle_2 + |y\rangle_1\,|x\rangle_2)\\
|\psi\rangle_{-} &= 2^{-1/2}(|x\rangle_1\,|y\rangle_2 - |y\rangle_1\,|x\rangle_2)\\
|\psi\rangle^{+} &= 2^{-1/2}(|x\rangle_1\,|x\rangle_2 + |y\rangle_1\,|y\rangle_2)\\
|\psi\rangle^{-} &= 2^{-1/2}(|x\rangle_1\,|x\rangle_2 - |y\rangle_1\,|y\rangle_2)
\end{aligned} \tag{6.60}$$

6.5 Quantum entanglement from the Pauli matrices

Pauli matrices are widely used in quantum mechanics. Here, an introduction to Pauli matrices via the dynamics of the Hamiltonian is given following Feynman's style (Feynman *et al* 1965) and the notation used by Duarte (2014). This introduction is followed by the derivation of the quantum entanglement probability amplitudes via the set of Pauli matrices.

6.5.1 Pauli matrices from the Hamiltonian

Pauli matrices can be introduced from the vantage point of the Hamiltonian as masterfully done by Feynman *et al* (1965). For a two-state system the time-dependent equation is (Dirac 1958)

$$i\hbar\frac{dC_i}{dt} = \sum_j H_{ij}C_j \tag{6.61}$$

which lead to the system

$$i\hbar\frac{dC_1}{dt} = H_{11}C_1 + H_{12}C_2 \tag{6.62}$$

$$i\hbar\frac{dC_2}{dt} = H_{21}C_1 + H_{22}C_2 \tag{6.63}$$

Under the influence of a magnetic field $\boldsymbol{B}$ the Hamiltonian, for a spin one-half particle, such as an electron, is given by (Feynman *et al* 1965)

$$\begin{pmatrix} H_{11} & H_{12} \\ H_{21} & H_{22} \end{pmatrix} = \begin{pmatrix} -\mu B_z & -\mu(B_x - iB_y) \\ -\mu(B_x + iB_y) & +\mu B_z \end{pmatrix} \tag{6.64}$$

In general, for a fermion, the Hamiltonian is defined as (Feynman *et al* 1965)

$$H_{ij} = -\mu\left(\sigma_{ij}^{x} B_x + \sigma_{ij}^{y} B_y + \sigma_{ij}^{z} B_z\right) \tag{6.65}$$

Based on the definitions given in equations (6.64) and (6.65), for the z component, that is B_z

$$H_{11} = -\mu\sigma_{11}^{z} B_z = -\mu B_z \tag{6.66}$$

$$H_{12} = -\mu\sigma_{12}^{z} B_z = 0 \tag{6.67}$$

$$H_{21} = -\mu\sigma_{21}^{z} B_z = 0 \tag{6.68}$$

$$H_{22} = -\mu\sigma_{22}^{z} B_z = +\mu B_z \tag{6.69}$$

which means

$$\sigma_{11}^{z} = 1 \tag{6.70}$$

$$\sigma_{12}^{z} = 0 \tag{6.71}$$

$$\sigma_{21}^{z} = 0 \tag{6.72}$$

$$\sigma_{22}^{z} = -1 \tag{6.73}$$

In matrix form (Dirac 1958)

$$\sigma_{ij}^{z} = \sigma_z = \begin{pmatrix} 1 & 0 \\ 0 & -1 \end{pmatrix} \tag{6.74}$$

Using the same methodology, for the x and y components (Dirac 1958)

$$\sigma_{ij}^{x} = \sigma_x = \begin{pmatrix} 0 & 1 \\ 1 & 0 \end{pmatrix} \tag{6.75}$$

$$\sigma_{ij}^{y} = \sigma_y = \begin{pmatrix} 0 & -i \\ i & 0 \end{pmatrix} \tag{6.76}$$

The 2×2 matrices σ_x, σ_y, and σ_z, are known as Pauli matrices (Dirac 1958) and are used widely in spin and magnetic moment computations. Furthermore, Pauli matrices are of great importance to quantum optics.

6.5.2 From Pauli matrices to quantum entanglement states

Previously it has been shown that a firm natural foundation for the quantum entanglement probability amplitudes can be provided by the Dirac–Feynman interferometric principle (Duarte 2013a, 2013b, 2014)

$$\langle d|s\rangle = \sum_{j=1}^{N} \langle d|j\rangle\langle j|s\rangle \tag{6.77}$$

Here it is shown that, assuming a rigorous and independent derivation of the Pauli matrices, then the equations for the probability amplitudes of quantum entanglement can be neatly derived using the vector form of the polarization states $|x\rangle$ and $|y\rangle$.

This section is based on the approach of Duarte *et al* (2020) and the notation utilized by Duarte (2019). Here, the following definitions are necessary: the $|x\rangle$ and $|y\rangle$ polarization states are defined in vector notation as

$$|x\rangle = \begin{pmatrix} 1 \\ 0 \end{pmatrix} \tag{6.78}$$

$$|y\rangle = \begin{pmatrix} 0 \\ 1 \end{pmatrix} \tag{6.79}$$

Further, the *direct vector product* (see chapter 5) is defined as (Ayres 1965)

$$|x\rangle|y\rangle = |x\rangle \cdot |y\rangle^T \tag{6.80}$$

which in explicit notation means

$$|x\rangle \cdot |y\rangle^T = \begin{pmatrix} x_1 \\ x_2 \end{pmatrix}\begin{pmatrix} y_1 \\ y_2 \end{pmatrix}^T = \begin{pmatrix} x_1y_1 & x_1y_2 \\ x_2y_1 & x_2y_2 \end{pmatrix} \tag{6.81}$$

Starting from the σ_x Pauli matrix given in equation (6.75)

$$\sigma_x = \begin{pmatrix} 0 & 1 \\ 1 & 0 \end{pmatrix} = \begin{pmatrix} 0 & 1 \\ 0 & 0 \end{pmatrix} + \begin{pmatrix} 0 & 0 \\ 1 & 0 \end{pmatrix} \tag{6.82}$$

$$\begin{pmatrix} 0 & 1 \\ 0 & 0 \end{pmatrix} + \begin{pmatrix} 0 & 0 \\ 1 & 0 \end{pmatrix} = \left(\begin{pmatrix} 1 \\ 0 \end{pmatrix}\begin{pmatrix} 0 \\ 1 \end{pmatrix} + \begin{pmatrix} 0 \\ 1 \end{pmatrix}\begin{pmatrix} 1 \\ 0 \end{pmatrix}\right) \tag{6.83}$$

$$\left(\begin{pmatrix} 1 \\ 0 \end{pmatrix}\begin{pmatrix} 0 \\ 1 \end{pmatrix} + \begin{pmatrix} 0 \\ 1 \end{pmatrix}\begin{pmatrix} 1 \\ 0 \end{pmatrix}\right) = (|x\rangle|y\rangle + |y\rangle|x\rangle) \tag{6.84}$$

which can be written, in terms of σ_x, as

$$2^{-1/2}(|x\rangle|y\rangle + |y\rangle|x\rangle) = 2^{-1/2}\sigma_x \tag{6.85}$$

Equation (6.85) can ultimately be expressed in probability amplitude notation as

$$|\psi\rangle_+ = 2^{-1/2}(|x\rangle|y\rangle + |y\rangle|x\rangle) \tag{6.86}$$

and

$$|\psi\rangle_+ = 2^{-1/2}\sigma_x \tag{6.87}$$

Considering the σ_y Pauli matrix given in equation (6.76)

$$\sigma_y = \begin{pmatrix} 0 & -i \\ i & 0 \end{pmatrix} = \begin{pmatrix} 0 & -i \\ 0 & 0 \end{pmatrix} + \begin{pmatrix} 0 & 0 \\ i & 0 \end{pmatrix} \tag{6.88}$$

$$-i\left(\begin{pmatrix}0 & 1\\0 & 0\end{pmatrix}-\begin{pmatrix}0 & 0\\1 & 0\end{pmatrix}\right)=-i\left(\begin{pmatrix}1\\0\end{pmatrix}\begin{pmatrix}0\\1\end{pmatrix}-\begin{pmatrix}0\\1\end{pmatrix}\begin{pmatrix}1\\0\end{pmatrix}\right) \tag{6.89}$$

$$-i\left(\begin{pmatrix}1\\0\end{pmatrix}\begin{pmatrix}0\\1\end{pmatrix}-\begin{pmatrix}0\\1\end{pmatrix}\begin{pmatrix}1\\0\end{pmatrix}\right)=-i(|x\rangle|y\rangle-|y\rangle|x\rangle) \tag{6.90}$$

which can be written, in terms of σ_y, as

$$2^{-1/2}(-i)(|x\rangle|y\rangle-|y\rangle|x\rangle)=2^{-1/2}i\sigma_y \tag{6.91}$$

Equation (6.91) can ultimately be expressed in probability amplitude notation as

$$|\psi\rangle_-=2^{-1/2}(|x\rangle|y\rangle-|y\rangle|x\rangle) \tag{6.92}$$

and

$$|\psi\rangle_-=2^{-1/2}i\sigma_y \tag{6.93}$$

Considering the σ_z Pauli matrix given in equation (6.74)

$$\sigma_z=\begin{pmatrix}1 & 0\\0 & -1\end{pmatrix}=\begin{pmatrix}1 & 0\\0 & 0\end{pmatrix}-\begin{pmatrix}0 & 0\\0 & 1\end{pmatrix} \tag{6.94}$$

$$\begin{pmatrix}1 & 0\\0 & 0\end{pmatrix}-\begin{pmatrix}0 & 0\\0 & 1\end{pmatrix}=\left(\begin{pmatrix}1\\0\end{pmatrix}\begin{pmatrix}1\\0\end{pmatrix}-\begin{pmatrix}0\\1\end{pmatrix}\begin{pmatrix}0\\1\end{pmatrix}\right) \tag{6.95}$$

$$\left(\begin{pmatrix}1\\0\end{pmatrix}\begin{pmatrix}1\\0\end{pmatrix}-\begin{pmatrix}0\\1\end{pmatrix}\begin{pmatrix}0\\1\end{pmatrix}\right)=(|x\rangle|x\rangle-|y\rangle|y\rangle) \tag{6.96}$$

which can be written, in terms of σ_z, as

$$2^{-1/2}(|x\rangle|x\rangle-|y\rangle|y\rangle)=2^{-1/2}\sigma_z \tag{6.97}$$

Equation (6.97) can ultimately be expressed in probability amplitude notation as

$$|\psi\rangle^-=2^{-1/2}(|x\rangle|x\rangle-|y\rangle|y\rangle) \tag{6.98}$$

and

$$|\psi\rangle^-=2^{-1/2}\sigma_z \tag{6.99}$$

Finally, considering the identity matrix I

$$I=\begin{pmatrix}1 & 0\\0 & 1\end{pmatrix} \tag{6.100}$$

it can be shown that

$$|\psi\rangle^+=2^{-1/2}(|x\rangle|x\rangle+|y\rangle|y\rangle) \tag{6.101}$$

$$|\psi\rangle^+=2^{-1/2}I \tag{6.102}$$

In summary, the probability amplitudes of quantum entanglement are related to the Pauli matrices and the identity matrix via the elegant set of identities (Duarte 2019, Duarte *et al* 2020)

$$\begin{aligned} |\psi\rangle_+ &= 2^{-1/2}\sigma_x \\ |\psi\rangle_- &= 2^{-1/2}i\sigma_y \\ |\psi\rangle^+ &= 2^{-1/2}I \\ |\psi\rangle^- &= 2^{-1/2}\sigma_z \end{aligned} \tag{6.103}$$

6.6 Quantum entanglement from the Hadamard gate

The Hadamard matrix, named after the mathematician J Hadamard, is a unitary matrix given by

$$H = \begin{pmatrix} 1 & 1 \\ 1 & -1 \end{pmatrix} \tag{6.104}$$

By inspection, it immediately becomes apparent that

$$H = (\sigma_x + \sigma_z) \tag{6.105}$$

Using equations (6.87) and (6.99) leads to

$$H = 2^{1/2}(|\psi\rangle_+ + |\psi\rangle^-) \tag{6.106}$$

Defining the Hadamard gate as

$$H_G = 2^{-1/2}\begin{pmatrix} 1 & 1 \\ 1 & -1 \end{pmatrix} \tag{6.107}$$

it follows that

$$H_G = (|\psi\rangle_+ + |\psi\rangle^-) \tag{6.108}$$

Thus the Hadamard matrix is related to two of the probability amplitudes of quantum entanglement $|\psi\rangle_+$ and $|\psi\rangle^-$. It can be shown that the other two probability amplitudes $|\psi\rangle^+$ and $|\psi\rangle_-$ are related to the Hadamard-like matrix K (Duarte and Taylor 2021)

$$K = \begin{pmatrix} 1 & 1 \\ -1 & 1 \end{pmatrix} \tag{6.109}$$

$$K = (I + i\sigma_y) \tag{6.110}$$

$$K_G = 2^{-1/2}\begin{pmatrix} 1 & 1 \\ -1 & 1 \end{pmatrix} \tag{6.111}$$

$$K_G = 2^{-1/2}(I + i\sigma_y) \tag{6.112}$$

$$K_G = (|\psi\rangle^+ + |\psi\rangle_-) \tag{6.113}$$

6.7 Quantum interference or quantum entanglement?

An important question which has been considered for a long time has to do with the fundamental nature of quantum entanglement. For instance, Schrödinger (1935, 1936) attached enormous importance, albeit in an interrogative mode, to the concept of quantum entanglement at the foundations of quantum mechanics. In addition Ward, in his terse style, wrote: '($|x, y\rangle - |y, x\rangle$) … *was my first lesson in quantum mechanics, and in a very real sense my last, since all the rest is mere technique, which can be learnt from books.*' (Ward 2004). On the other hand, another quantum master, Feynman, considered interference to be at the *heart* of quantum mechanics (Feynman *et al* 1965). In this chapter an interplay between quantum interference and quantum entanglement has been exhibited in detail. The question to be now considered is: which principle is more fundamental in quantum mechanics: quantum entanglement or quantum interference?

The material considered here is an extension of the discussion already provided by Duarte (2019).

Assuming that the Dirac–Feynman interferometric principle

$$\langle d|s\rangle = \sum_{j=1}^{N} \langle d|j\rangle\langle j|s\rangle$$

is an intrinsic foundational principle of quantum mechanics it can be applied to derive the generalized principles of quantum entanglement (Duarte 2016, 2019, Duarte and Taylor 2017, 2021)

$$|\psi\rangle_R = N^{-1/2}\sum_{j=1}^{N}(\pm)|C\rangle_{N+1-j} \tag{6.114}$$

$$1 = ||\psi\rangle_I|^2 + ||\psi\rangle_{II}|^2 + ||\psi\rangle_{III}|^2 + ||\psi\rangle_{IV}|^2 + \ldots \tag{6.115}$$

$$|C\rangle_{N+1-j} = \prod_{m=1,3,5\ldots}^{n} |a\rangle_m\, |b\rangle_{m+1} \tag{6.116}$$

which for a pair of quanta and two propagation channels ($n = N = 2$) become

$$|\psi\rangle = 2^{-1/2}(|x\rangle_1\, |y\rangle_2 \pm |y\rangle_1\, |x\rangle_2) \tag{6.117}$$

In this regard, the interferometric path to quantum entanglement is natural, clear, and transparent.

Given the reversible nature of the quantum mechanical process, the avenue

$$2^{-1/2}(|x\rangle_1\, |y\rangle_2 \pm |y\rangle_1\, |x\rangle_2) \rightarrow \sum_{j=1}^{N}\langle d|j\rangle\langle j|s\rangle \tag{6.118}$$

is also clear and transparent. However, the derivation of the quantum entanglement probability amplitude via non interferometric physics, at this stage, appears to be somewhat removed from foundational physics principles and more related to mathematical procedures involving the Pauli matrices or the Hahamard matrix.

The interferometric route to quantum entanglement is rather direct

$$\sum_{j=1}^{N} \langle d|j\rangle\langle j|s\rangle \rightarrow 2^{-1/2}(|x\rangle_1 |y\rangle_2 \pm |y\rangle_1 |x\rangle_2) \tag{6.119}$$

with the main intermediate step involving the Dirac identities $|C\rangle_2 = |x\rangle_1 |x\rangle_2$ and $|C\rangle_1 = |y\rangle_1 |y\rangle_2$ (Duarte 2013a, 2014, 2016). On the other hand, in addition to the use of the Dirac identities, the derivational path from quantum entanglement to interference requires

$$\sum_{j} H_{ij}C_j \rightarrow \sigma_x, \sigma_y \rightarrow 2^{-1/2}(|x\rangle_1 |y\rangle_2 \pm |y\rangle_1 |x\rangle_2) \rightarrow \sum_{j=1}^{N} \langle d|j\rangle\langle j|s\rangle \tag{6.120}$$

which appears to imply that quantum interference is a more fundamental principle.

From an entirely neutral corner, assuming that both

$$\langle d|s\rangle = \sum_{j=1}^{N} \langle d|j\rangle\langle j|s\rangle$$

and

$$|\psi\rangle = 2^{-1/2}(|x\rangle_1 |y\rangle_2 \pm |y\rangle_1 |x\rangle_2)$$

are *intrinsic foundational principles* then it can be deduced that these two principles are strongly interconnected at the most basic level.

Finally, it should be emphasized that neither Dirac, nor Feynman, ever produced a proof for what we now call the Dirac–Feynman principle. The first hint of a 'proof' for this principle is via the reversibility (Duarte 2019)

$$2^{-1/2}(|x\rangle_1 |y\rangle_2 \pm |y\rangle_1 |x\rangle_2) \rightarrow \sum_{j=1}^{N} \langle d|j\rangle\langle j|s\rangle$$

Problems

6.1 Show that equation (6.19) follows from equation (6.18).

6.2 Starting from equation (6.64), verify that $\sigma_{ij}^{x} = \sigma_x = \begin{pmatrix} 0 & 1 \\ 1 & 0 \end{pmatrix}$.

6.3 Starting from equation (6.64), verify that $\sigma_{ij}^{y} = \sigma_y = \begin{pmatrix} 0 & -i \\ i & 0 \end{pmatrix}$.

6.4 Starting from equation (6.100) show that $|\psi\rangle^{+} = 2^{-1/2}(|x\rangle|x\rangle + |y\rangle|y\rangle)$.

6.5 Show that $|\psi\rangle^{+} = 2^{-1/2}I$.

6.6 Show that $H_G = (|\psi\rangle_{+} + |\psi\rangle^{-})$.

6.7 Starting from equation (6.111) show that $K_G = (|\psi\rangle^{+} + |\psi\rangle_{-})$.

References

Ayres F 1965 *Modern Algebra* (New York: McGraw-Hill)

Dirac P A M 1958 *The Principles of Quantum Mechanics* (Oxford: Oxford University Press)

Duarte F J 2013a The probability amplitude for entangled polarizations: an interferometric approach *J. Mod. Opt.* **60** 1585–7

Duarte F J 2013b Tunable laser optics: applications to optics and quantum optics *Prog. Quantum Electron.* **37** 326–47

Duarte F J 2014 *Quantum Optics for Engineers* (New York: CRC Press)

Duarte F J 2016 Secure space-to-space interferometric communications and its nexus to the physics of quantum entanglement *Appl. Phys. Rev.* **3** 041301

Duarte F J 2019 *Fundamentals of Quantum Entanglement* (Bristol: IOP Publishing)

Duarte F J and Taylor T S 2017 Quantum entanglement probability amplitudes in multiple channels: an interferometric approach *Optik* **139** 222–30

Duarte F J and Taylor T S 2021 *To be published*

Duarte F J, Taylor T S and Slaten J C 2020 On the probability amplitude of quantum entanglement and the Pauli matrices *Opt. Quantum Electron.* **52** 106

Feynman R P, Leighton R B and Sands M 1965 *The Feynman Lectures on Physics* vol III (Reading, MA: Addison-Wesley)

Mandel L and Wolf E 1995 *Optical Coherence and Quantum Optics* (Cambridge: Cambridge University Press)

Pryce M L H and Ward J C 1947 Angular correlation effects with annihilation radiation *Nature* **160** 435

Schrödinger E 1935 Discussion of probability relations between separated systems *Math. Proc. Camb. Phil. Soc.* **31** 555–63

Schrödinger E 1936 Probability relations between separated systems *Math. Proc. Camb. Phil. Soc.* **32** 446–52

Ward J C 1949 *Some Properties of the Elementary Particles* (Oxford: Oxford University Press)

Ward J C 2004 *Memoirs of a Theoretical Physicist* (Rochester, NY: Optics Journal)

Wheeler J A 1946 Polyelectrons *Ann. New York Acad. Sci.* **48** 219–38

Chapter 7

Quantum entanglement in matrix notation

The probability amplitudes for quantum entanglement for $n = N = 2$ are expressed in 2×2 matrix form. The explicit mathematical nexus between the probability amplitudes for quantum entanglement with the Pauli matrices and the Hadamard matrix is described. From this approach, a series of new 2×2 optical matrices are introduced.

7.1 Introduction

In this chapter the complete set of probability amplitudes for quantum entanglement, also known as Bell states, is expressed in 2×2 matrix form. These are 2×2 matrices that provide information about the polarization of the quanta, that is $|x\rangle$ and $|y\rangle$. Particular attention is given to the nexus between the probability amplitudes for quantum entanglement and the Pauli matrices as well as the link between probability amplitudes for quantum entanglement and the Hadamard matrix. A new set of Hadamard-type matrices is introduced and their properties discussed.

7.2 Quantum entanglement probability amplitudes

The probability amplitude for two quanta propagating in opposite directions, with entangled orthogonal polarizations $|x\rangle$ and $|y\rangle$, is given by (Pryce and Ward 1947, Duarte 2013, 2014)

$$|\psi\rangle_{\pm} = 2^{-1/2}(|x\rangle|y\rangle \pm |y\rangle|x\rangle) \tag{7.1}$$

The complete set of probability amplitudes also include the following alternatives (Duarte 2019)

$$|\psi\rangle^{\pm} = 2^{-1/2}(|x\rangle|x\rangle \pm |y\rangle|y\rangle) \tag{7.2}$$

doi:10.1088/978-0-7503-3407-5ch7

Defining the orthogonal polarization states as $|x\rangle = |1\rangle$ and $|y\rangle = |0\rangle$, the complete set of probability amplitude for quantum entanglement can be expressed as (Duarte 2019)

$$|\psi\rangle_+ = 2^{-1/2}(|1\rangle|0\rangle + |0\rangle|1\rangle) \tag{7.3}$$

$$|\psi\rangle_- = 2^{-1/2}(|1\rangle|0\rangle - |0\rangle|1\rangle) \tag{7.4}$$

$$|\psi\rangle^+ = 2^{-1/2}(|1\rangle|1\rangle + |0\rangle|0\rangle) \tag{7.5}$$

$$|\psi\rangle^- = 2^{-1/2}(|1\rangle|1\rangle - |0\rangle|0\rangle) \tag{7.6}$$

7.3 From *ket* vectors to polarization matrices

As indicated previously (Duarte 2019, Duarte *et al* 2020) the quantum entanglement probability amplitudes can also be expressed in matrix form using the *vector direct product* (see chapter 5) defined as (Ayres 1965)

$$|x\rangle|y\rangle = |x\rangle \cdot |y\rangle^T \tag{7.7}$$

$$|x\rangle \cdot |y\rangle^T = \begin{pmatrix} x_1 \\ x_2 \end{pmatrix}\begin{pmatrix} y_1 \\ y_2 \end{pmatrix}^T = \begin{pmatrix} x_1y_1 & x_1y_2 \\ x_2y_1 & x_2y_2 \end{pmatrix} \tag{7.8}$$

Defining the *ket* vectors $|x\rangle$ and $|y\rangle$ as

$$|x\rangle = |1\rangle = \begin{pmatrix} 1 \\ 0 \end{pmatrix} \tag{7.9}$$

$$|y\rangle = |0\rangle = \begin{pmatrix} 0 \\ 1 \end{pmatrix} \tag{7.10}$$

application of the vector direct product to the set of probability amplitudes for quantum entanglement leads to

$$|\psi\rangle_+ = 2^{-1/2}\left(\begin{pmatrix} 1 \\ 0 \end{pmatrix} \cdot \begin{pmatrix} 0 \\ 1 \end{pmatrix} + \begin{pmatrix} 0 \\ 1 \end{pmatrix} \cdot \begin{pmatrix} 1 \\ 0 \end{pmatrix}\right) = 2^{-1/2}\left(\begin{pmatrix} 0 & 1 \\ 0 & 0 \end{pmatrix} + \begin{pmatrix} 0 & 0 \\ 1 & 0 \end{pmatrix}\right) \tag{7.11}$$

$$|\psi\rangle_- = 2^{-1/2}\left(\begin{pmatrix} 1 \\ 0 \end{pmatrix} \cdot \begin{pmatrix} 0 \\ 1 \end{pmatrix} - \begin{pmatrix} 0 \\ 1 \end{pmatrix} \cdot \begin{pmatrix} 1 \\ 0 \end{pmatrix}\right) = 2^{-1/2}\left(\begin{pmatrix} 0 & 1 \\ 0 & 0 \end{pmatrix} - \begin{pmatrix} 0 & 0 \\ 1 & 0 \end{pmatrix}\right) \tag{7.12}$$

$$|\psi\rangle^+ = 2^{-1/2}\left(\begin{pmatrix} 1 \\ 0 \end{pmatrix} \cdot \begin{pmatrix} 1 \\ 0 \end{pmatrix} + \begin{pmatrix} 0 \\ 1 \end{pmatrix} \cdot \begin{pmatrix} 0 \\ 1 \end{pmatrix}\right) = 2^{-1/2}\left(\begin{pmatrix} 1 & 0 \\ 0 & 0 \end{pmatrix} + \begin{pmatrix} 0 & 0 \\ 0 & 1 \end{pmatrix}\right) \tag{7.13}$$

$$|\psi\rangle^- = 2^{-1/2}\left(\begin{pmatrix} 1 \\ 0 \end{pmatrix} \cdot \begin{pmatrix} 1 \\ 0 \end{pmatrix} - \begin{pmatrix} 0 \\ 1 \end{pmatrix} \cdot \begin{pmatrix} 0 \\ 1 \end{pmatrix}\right) = 2^{-1/2}\left(\begin{pmatrix} 1 & 0 \\ 0 & 0 \end{pmatrix} - \begin{pmatrix} 0 & 0 \\ 0 & 1 \end{pmatrix}\right) \tag{7.14}$$

These equations are equivalent to

$$|\psi\rangle_{+} = 2^{-1/2}\begin{pmatrix} 0 & 1 \\ 1 & 0 \end{pmatrix} \tag{7.15}$$

$$|\psi\rangle_{-} = 2^{-1/2}i\begin{pmatrix} 0 & -i \\ i & 0 \end{pmatrix} \tag{7.16}$$

$$|\psi\rangle^{+} = 2^{-1/2}\begin{pmatrix} 1 & 0 \\ 0 & 1 \end{pmatrix} \tag{7.17}$$

$$|\psi\rangle^{-} = 2^{-1/2}\begin{pmatrix} 1 & 0 \\ 0 & -1 \end{pmatrix} \tag{7.18}$$

7.4 The Pauli matrices and quantum entanglement

A succinct manner to express equations (7.15)–(7.18) is (Duarte *et al* 2020)

$$|\psi\rangle_{+} = 2^{-1/2}\,\sigma_x \tag{7.19}$$

$$|\psi\rangle_{-} = 2^{-1/2}\,i\sigma_y \tag{7.20}$$

$$|\psi\rangle^{+} = 2^{-1/2}\,I \tag{7.21}$$

$$|\psi\rangle^{-} = 2^{-1/2}\,\sigma_z \tag{7.22}$$

where

$$\sigma_x = \begin{pmatrix} 0 & 1 \\ 1 & 0 \end{pmatrix} \tag{7.23}$$

$$\sigma_y = \begin{pmatrix} 0 & -i \\ i & 0 \end{pmatrix} \tag{7.24}$$

$$\sigma_z = \begin{pmatrix} 1 & 0 \\ 0 & -1 \end{pmatrix} \tag{7.25}$$

$$I = \begin{pmatrix} 1 & 0 \\ 0 & 1 \end{pmatrix} \tag{7.26}$$

are the σ_x, σ_y, σ_z Pauli matrices and I is the identity matrix (Dirac 1958). See chapter 6 for a discussion on the physics of the Pauli matrices.

7.5 The Hadamard matrix

The Hadamard matrix is an important matrix to the fields of quantum computing and, as will be seen here, to quantum optics itself. This matrix is named after the mathematician J Hadamard and is given by

$$H = \begin{pmatrix} 1 & 1 \\ 1 & -1 \end{pmatrix} \tag{7.27}$$

Equation (7.27) can be re expressed as the sum of two matrices

$$H = \begin{pmatrix} 0 & 1 \\ 1 & 0 \end{pmatrix} + \begin{pmatrix} 1 & 0 \\ 0 & -1 \end{pmatrix} \tag{7.28}$$

which, by inspection, can also be written as (Duarte 2019)

$$H = \sigma_x + \sigma_z \tag{7.29}$$

7.5.1 Derivation of the matrix for the Hadamard gate

Considering the quantum entanglement states given in equation (7.1)

$$|\psi\rangle_+ = 2^{-1/2}(|x\rangle_1|y\rangle_2 + |y\rangle_1|x\rangle_2) \tag{7.30}$$

$$|\psi\rangle_- = 2^{-1/2}(|x\rangle_1|y\rangle_2 - |y\rangle_1|x\rangle_2) \tag{7.31}$$

one can write (Dirac 1958, Duarte 2013)

$$|\alpha\rangle = |x\rangle_1|y\rangle_2 \tag{7.32}$$

$$|\beta\rangle = |y\rangle_1|x\rangle_2 \tag{7.33}$$

which leads directly to the equations describing beam splitter based interferometers such as the Mach–Zehnder interferometer (Duarte 2003, 2014, 2019)

$$|\Psi\rangle_+ = 2^{-1/2}(|\alpha\rangle + |\beta\rangle) \tag{7.34}$$

$$|\Psi\rangle_- = 2^{-1/2}(|\alpha\rangle - |\beta\rangle) \tag{7.35}$$

Notice that these equations have the exact form of the equations derived directly from the physics of the Mach–Zehnder interferometer, given in chapter 4, relative to the output beam splitter with two orthogonal inputs (Duarte 2014, 2019).

Equations (7.34) and (7.35) can next be expressed in matrix form as

$$\begin{pmatrix} |\Psi\rangle_+ \\ |\Psi\rangle_- \end{pmatrix} = 2^{-1/2}\begin{pmatrix} 1 & 1 \\ 1 & -1 \end{pmatrix}\begin{pmatrix} |\alpha\rangle \\ |\beta\rangle \end{pmatrix} \tag{7.36}$$

where

$$2^{-1/2}\begin{pmatrix} 1 & 1 \\ 1 & -1 \end{pmatrix} = H_G \tag{7.37}$$

is the Hadamard gate H_G widely used in the literature in the absence of a transparent derivation.

7.6 Optical matrices based on the probability amplitudes of quantum entanglement

The Hadamard gate H_G, derived in the previous section, can also be defined via the multiplication of the Hadamard matrix by the normalization factor $2^{-1/2}$, so that

$$H_G = 2^{-1/2}\begin{pmatrix} 1 & 1 \\ 1 & -1 \end{pmatrix}$$

which can also be expressed in terms of the probability amplitudes for quantum entanglement as (Duarte *et al* 2020)

$$H_G = (|\psi\rangle_+ + |\psi\rangle^-) \tag{7.38}$$

$$H_G = 2^{-1/2}(\sigma_x + \sigma_z) \tag{7.39}$$

Here, new matrices become immediately apparent (Duarte and Taylor 2020)

$$J = (|\psi\rangle_+ - |\psi\rangle^-) \tag{7.40}$$

$$J = 2^{-1/2}(\sigma_x - \sigma_z) \tag{7.41}$$

$$J = 2^{-1/2}\begin{pmatrix} -1 & 1 \\ 1 & 1 \end{pmatrix} \tag{7.42}$$

$$K = (|\psi\rangle^+ + |\psi\rangle_-) \tag{7.43}$$

$$K = 2^{-1/2}(I + i\sigma_y) \tag{7.44}$$

$$K = 2^{-1/2}\begin{pmatrix} 1 & 1 \\ -1 & 1 \end{pmatrix} \tag{7.45}$$

$$L = (|\psi\rangle^+ - |\psi\rangle_-) \tag{7.46}$$

$$L = 2^{-1/2}(I - i\sigma_y) \tag{7.47}$$

$$L = 2^{-1/2}\begin{pmatrix} 1 & -1 \\ 1 & 1 \end{pmatrix} \tag{7.48}$$

$$M = (|\psi\rangle^+ + |\psi\rangle_+) \tag{7.49}$$

$$M = 2^{-1/2}(I + \sigma_x) \tag{7.50}$$

$$M = 2^{-1/2}\begin{pmatrix} 1 & 1 \\ 1 & 1 \end{pmatrix} \tag{7.51}$$

$$N = (|\psi\rangle^+ - |\psi\rangle_+) \tag{7.52}$$

$$N = 2^{-1/2}(I - \sigma_x) \quad (7.53)$$

$$N = 2^{-1/2}\begin{pmatrix} 1 & -1 \\ -1 & 1 \end{pmatrix} \quad (7.54)$$

$$O = (|\psi\rangle^{-} + |\psi\rangle_{-}) \quad (7.55)$$

$$O = 2^{-1/2}(\sigma_z + i\sigma_y) \quad (7.56)$$

$$O = 2^{-1/2}\begin{pmatrix} 1 & 1 \\ -1 & -1 \end{pmatrix} \quad (7.57)$$

$$P = (|\psi\rangle^{-} - |\psi\rangle_{-}) \quad (7.58)$$

$$P = 2^{-1/2}(\sigma_z - i\sigma_y) \quad (7.59)$$

$$P = 2^{-1/2}\begin{pmatrix} 1 & -1 \\ 1 & -1 \end{pmatrix} \quad (7.60)$$

Here it should be noted that $M = (|\psi\rangle^{+} + |\psi\rangle_{+})$ can be configured with a combination of an ordinary beam splitter and a polarization rotator (Duarte and Taylor 2020).

7.7 Polarization rotators for quantum entanglement

The physics of polarization rotators is discussed by Robson (1974) and Born and Wolf (1999). A more experimental perspective is given by Duarte (2014, 2015). In this section the matrix representation for wave plates, rhomboids, and multiple-prism rotators is presented. The operation of $\theta = \pi/2$ polarization rotators on quantum states is also discussed.

7.7.1 Wave plates

The generalized matrix for birefringent rotators is given by (Robson 1974)

$$\begin{pmatrix} \cos\theta & \sin\theta \\ -e^{i\delta}\sin\theta & e^{i\delta}\cos\theta \end{pmatrix} \quad (7.61)$$

For a quarter-wave plate $\delta = \pi/2$, the phase term is $e^{i\pi/2} = +i$, so that

$$R_{1/4} = \begin{pmatrix} \cos\theta & \sin\theta \\ -i\sin\theta & i\cos\theta \end{pmatrix} \quad (7.62)$$

For a half-wave plate $\delta = \pi$ and $e^{i\pi} = -1$, so that

$$R_{1/2} = \begin{pmatrix} \cos\theta & \sin\theta \\ \sin\theta & -\cos\theta \end{pmatrix} \quad (7.63)$$

Half-wave plates cause rotation of linearly polarized beams by $\theta = \pi/2$ so that the rotation matrix reduces to

$$R_{1/2} = \begin{pmatrix} 0 & 1 \\ 1 & 0 \end{pmatrix} \tag{7.64}$$

Quarter-wave plates and half-wave are wavelength specific. However, their wavelength performance can be improved using achromatic designs incorporating multiple crystal materials.

7.7.2 Rhomboids

Another useful polarization rotator is the double Fresnel rhomb, which illustrated in figure 7.1. This polarization rotator is wavelength specific and it rotates polarized light by $\theta = \pi/2$ so that its polarization matrix P_R is

$$R_R = \begin{pmatrix} 0 & 1 \\ 1 & 0 \end{pmatrix} \tag{7.65}$$

7.7.3 Prismatic rotators

One further polarization rotator is the broadband collinear multiple-prism rotator (Duarte 1989) displayed in figure 7.2. It also rotates linearly polarized light by $\theta = \pi/2$ so that its polarization matrix R_{MP} is given by

$$R_{MP} = \begin{pmatrix} 0 & 1 \\ 1 & 0 \end{pmatrix} \tag{7.66}$$

This broadband collinear multiple-prism rotator has demonstrated high-fidelity transmission, for $\theta = \pi/2$ rotation, at efficiencies approaching 95% (Duarte 1992). Also, since the prismatic configuration can be held in place mechanically, thus

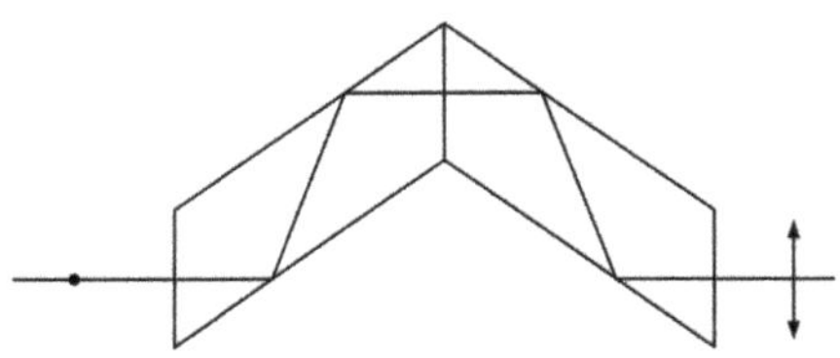

Figure 7.1. Schematics for a generic wavelength specific double Fresnel rhomb utilized for $\theta = \pi/2$ rotation of linearly polarized light (Duarte 2019).

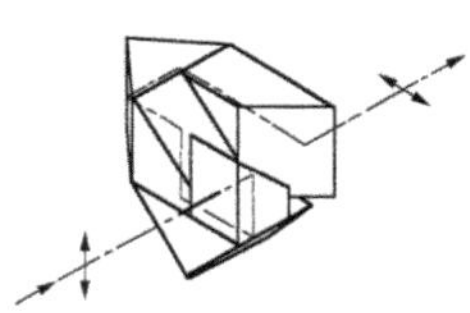

Figure 7.2. Isometric schematics of the broadband collinear multiple-prism rotator utilized for $\theta = \pi/2$ rotation of linearly polarized light (from Duarte 1989).

offering only air-to-glass interfaces, it is particularly apt to rotate the polarization of high-power laser beams.

7.8 Quantum operations with polarization rotators

If the entangled photons are made to be incident on a generalized $\theta = \pi/2$ polarization rotator represented by the matrix

$$R = \begin{pmatrix} 0 & 1 \\ 1 & 0 \end{pmatrix}$$

then it can be shown that (Duarte *et al* 2020)

$$R|\psi\rangle^{+} = |\psi\rangle_{-} = 2^{-1/2}\,\sigma_x \tag{7.67}$$

$$R|\psi\rangle^{-} = -\,|\psi\rangle_{-} = -2^{-1/2}\,i\sigma_y \tag{7.68}$$

$$R\,|\psi\rangle_{+} = |\psi\rangle^{+} = 2^{-1/2}\,I \tag{7.69}$$

$$R|\psi\rangle_{-} = -\,|\psi\rangle^{-} = -2^{-1/2}\,\sigma_z \tag{7.70}$$

This demonstrates the inherent capability for mathematical operations utilizing polarization rotators and entangled states. This simple concept means that beams of entangled photons interacting with optical apparata configured with optical elements capable of transforming the entangled states can be used to perform optical quantum computations.

7.9 Quantum operations with the Hadamard gate

It can also be shown that the Hadamard matrix operation on the probability amplitudes for quantum entanglement yield the following mathematical identities (Duarte *et al* 2020)

$$H_G|\psi\rangle_{+} = 2^{-1/2}(|\psi\rangle^{+} + |\psi\rangle_{-}) \tag{7.71}$$

$$H_G|\psi\rangle_{-} = 2^{-1/2}(|\psi\rangle_{+} - |\psi\rangle^{-}) \tag{7.72}$$

$$H_G|\psi\rangle^{+} = 2^{-1/2}(|\psi\rangle_{+} + |\psi\rangle^{-}) \tag{7.73}$$

$$H_G|\psi\rangle^{-} = 2^{-1/2}(|\psi\rangle^{+} - |\psi\rangle_{-}) \tag{7.74}$$

which are equivalent to (Duarte *et al* 2020)

$$H_G|\psi\rangle_{-} = 2^{-1/2}(I + i\sigma_y) \tag{7.75}$$

$$H_G|\psi\rangle_{-} = 2^{-1}(\sigma_x - \sigma_z) \tag{7.76}$$

$$H_G|\psi\rangle^{+} = 2^{-1}(\sigma_x + \sigma_z) \tag{7.77}$$

$$H_G|\psi\rangle^{-} = 2^{-1}(I - i\sigma_y) \tag{7.78}$$

Problems

7.1 Starting from equation (7.12), verify that $|\psi\rangle_- = 2^{-1/2}\, i\sigma_y$.
7.2 Starting from equation (7.14), verify that $|\psi\rangle^- = 2^{-1/2}\, \sigma_z$.
7.3 Show that $H_G = (\psi_+ + \psi^-)$.
7.4 Show that $J = (|\psi\rangle_+ - |\psi\rangle^-)$ and $J = 2^{-1/2}(\sigma_x - \sigma_z)$ are equivalent.
7.5 Show that $M = (|\psi\rangle^+ + |\psi\rangle_+)$ leads to equation (7.51).
7.6 Show that $O = (|\psi\rangle^- + |\psi\rangle_-)$ and $O = 2^{-1/2}(\sigma_z + i\sigma_y)$ are equivalent.
7.7 Show that $P = (|\psi\rangle^- - |\psi\rangle_-)$ leads to equation (7.60).
7.8 Show that $R|\psi\rangle^+ = |\psi\rangle_-$ is equivalent to $R|\psi\rangle^+ = 2^{-1/2}\, \sigma_x$.
7.9 Show that $H_G|\psi\rangle_+ = 2^{-1/2}(|\psi\rangle^+ + |\psi\rangle_-)$.
7.10 Show that $H_G|\psi\rangle^- = 2^{-1}(I - i\sigma_y)$.

References

Ayres F 1965 *Modern Algebra* (New York: McGraw-Hill)

Born M and Wolf E 1999 *Principles of Optics* (Cambridge: Cambridge University Press)

Dirac P A M 1958 *The Principles of Quantum Mechanics* 4th edn (Oxford: Oxford University Press)

Duarte F J 1989 Optical device for rotating the polarization of a light beam *US Patent* 4822150

Duarte F J 1992 Beam transmission characteristics of a collinear polarization rotator *Appl. Opt.* **31** 3377–8

Duarte F J 2003 *Tunable Laser Optics* (New York: Elsevier Academic)

Duarte F J 2013 The probability amplitude for entangled polarizations: an interferometric approach *J. Mod. Opt.* **60** 1585–7

Duarte F J 2014 *Quantum Optics for Engineers* (New York: CRC Press)

Duarte F J 2015 *Tunable Laser Optics* 2nd edn (New York: CRC Press)

Duarte F J 2019 *Fundamentals of Quantum Entanglement* (Bristol: IOP Publishing)

Duarte F J and Taylor T S 2021 Quantum entanglement in matrix notation: application to the Mach–Zehnder interferometer *To be published*

Duarte F J, Taylor T S and Slaten J C 2020 On the probability amplitude of quantum entanglement and the Pauli matrices *Opt. Quantum Electron.* **52** 106

Pryce M L H and Ward J C 1947 Angular correlation effects with annihilation radiation *Nature* **160** 435

Robson B A 1974 *The Theory of Polarization Phenomena* (Oxford: Clarendon)

IOP Publishing

Quantum Entanglement Engineering and Applications

F J Duarte and T S Taylor

Chapter 8

Quantum entanglement applications

The application of the probability amplitude for quantum entanglement, $|\psi\rangle = (|x\rangle_1|y\rangle_2 \pm |y\rangle_1|x\rangle_2)$, to quantum cryptography communications, quantum teleportation, quantum computing, and quantum metrology is described. An overview of additional applications is also provided.

8.1 Introduction

Three of the main applications of quantum entanglement probability amplitudes are *quantum cryptography*, *quantum teleportation*, and *quantum computing*. In this chapter attention is focused on these three standard applications as well as other additional applications of interest such as quantum metrology.

8.2 Classical cryptography concepts

Cryptography is an ancient word adopted from the Greek *κρυπτος*. The word cryptography integrates the concept of *hidden* and the concept of *writing*, or *secret writing*.

Throughout the ages, cryptography has depended on a *secret key* or *code* for its effectiveness.

This key, or code, is known to the two participants in the communication arrangement.

The plain language method is coded, transmitted, and decoded. The main vulnerability in this methodology is precisely the key, or code, that can be captured and deciphered or decrypted by a third party, spy, eavesdropper, or enemy.

During the American Revolutionary War, George Washington utilized a simple methodology to encrypt messages. In this key, numbers were associated with words, for instance: 1 meant *a*, 2 meant *an*, 3 meant *all*. With this simple technique hundreds of relevant words were encoded and secure messages were successfully sent and received.

doi:10.1088/978-0-7503-3407-5ch8

A more flexible and versatile methodology is to associate a definite symbol to a particular alpha-numerical character. This is demonstrated in figure 8.1 encoding the speed of light $c = 2.997\ 924\ 58 \times 10^8$ m s^{-1}. A widely used system of cryptographic communications involves *symmetric key distribution* and is illustrated in figure 8.2. In this approach the message is encrypted using a previously conceived key, which is made accessible in a secure manner to both the sender and the receiver, and transmitted via an unsecured channel. Upon arrival the message is deciphered by the receiver utilizing the previously established code. In contemporaneous settings symmetric key algorithms are employed to perform encryption and decryption.

An alternative methodology to the symmetric approach is to employ *public key cryptography*. This method utilizes both a public key and a secured key, which are mathematically linked. One key encrypts the alpha-numerical message and the other key deciphers the encrypted message back to the original alpha-numerical text (see figure 8.3).

The disadvantage of classical methods is that they are vulnerable to third party intrusion, code theft by spies, and clever computerized deciphering methods.

8.3 Quantum entanglement applications to cryptography

Quantum cryptography is the application of the superposition probability amplitude of quantum entanglement to enable the secure creation of a quantum key, or quantum code.

Once a quantum key has been established, communications proceed via conventional cryptographic methods.

According to Bennett (1992), *quantum key distribution* (QKD) is 'the classically impossible task of distributing secret information over an insecure channel whose transmissions are subject to inspection by an eavesdropper'.

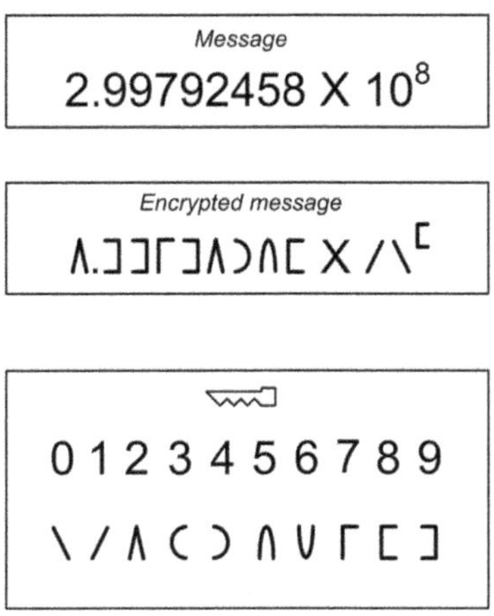

Figure 8.1. Symbolic code representing digits 0, 1, 2, 3, 4, 5, 6, 7, 8, 9 and symbolic. representation of the numerical value of the speed of light $c = 2.997\ 924\ 58 \times 10^8$ m s^{-1} using the code introduced in Duarte (2014).

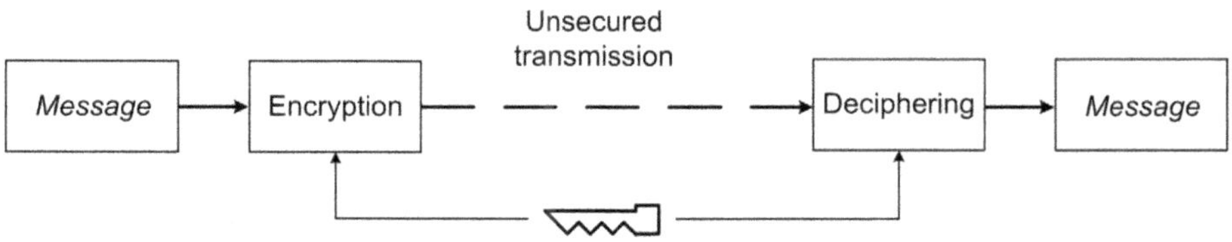

Figure 8.2. Schematics for classical symmetrical key cryptography. The key, or code, is identical for the sender and the receiver.

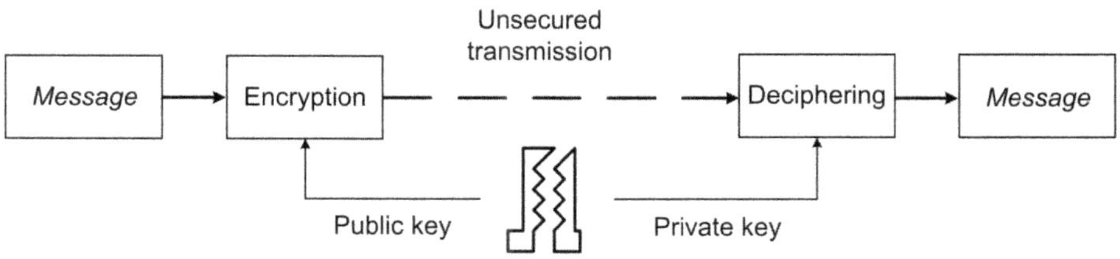

Figure 8.3. Schematics for classical public key cryptography. Two distinct keys, or codes, are used. One for the sender, which is public, and one for the receiver, which is private.

Quantum cryptography was first conceptually considered by Wiesner (1983). A scheme to implement quantum cryptography was introduced by Bennett and Brassard (1984) and further developed by Bennett *et al* (1992). The Bennett–Brassard method is based on the utilization of the polarization properties of propagating quanta utilizing nomenclature such as

$$|H\rangle = |x\rangle \tag{8.1}$$

$$|V\rangle = |y\rangle \tag{8.2}$$

$$|L\rangle = 2^{-1/2}(|x\rangle - i|y\rangle) \tag{8.3}$$

$$|R\rangle = 2^{-1/2}(|x\rangle + i|y\rangle) \tag{8.4}$$

where $|x\rangle$ is the state associated with x-polarized photons and $|y\rangle$ is the state associated with y-polarized photons. Moreover, $|L\rangle$ is the state associated to left-circularly polarized photons and $|R\rangle$ is the state associated to right-circularly polarized photons.

The Bennett and Brassard approach is described in detail by Bennett (1992), Bennett and Brassard (2014), and Bennett *et al* (1992, 1993). A brief pragmatic review is given by Duarte (2014).

In this section attention is given exclusively to free-space quantum cryptography, based on the physics of the quantum entanglement superposition probability amplitude.

A large area of research activity includes the methods of encoding, such as polarization-coding and phase-coding. For a review on those methods the reader should consult the works of Gisin *et al* (2002) and Casado *et al* (2008).

A persistent feature of conventional quantum cryptography based on Bell's theorem is that it requires a *key* or *code* to be created and distributed. In this regard, the reader should be informed that secure quantum communication techniques based on N-slit quantum interferometry and ensemble of indistinguishable photons, do not require key distribution (Duarte 2002, 2019).

8.3.1 Quantum key distribution (QKD)

The complete set of quantum entanglement probability amplitudes (Pryce and Ward 1947, Ward 1949) is given by (see chapter 2)

$$|\psi\rangle_+ = 2^{-1/2}(|x\rangle_1|y\rangle_2 + |y\rangle_1|x\rangle_2) \tag{8.5}$$

$$|\psi\rangle_- = 2^{-1/2}(|x\rangle_1|y\rangle_2 - |y\rangle_1|x\rangle_2) \tag{8.6}$$

$$|\psi\rangle^+ = 2^{-1/2}(|x\rangle_1|x\rangle_2 + |y\rangle_1|y\rangle_2) \tag{8.7}$$

$$|\psi\rangle^- = 2^{-1/2}(|x\rangle_1|x\rangle_2 - |y\rangle_1|y\rangle_2) \tag{8.8}$$

From this set the probability amplitudes given in equations (8.5) and (8.6) were ***implicitly*** introduced to the field of quantum cryptography by Ekert (1991). This author described a cryptographic methodology applicable to fermions. The technique is based on Bell's theorem (Bell 1964) via the Bohm-EPR perspective (Einstein *et al* 1935, Bohm 1951). It should be noted that the use of Bell's inequalities implies the calculation of quantum probabilities using the probability amplitudes given in equations (8.5)–(8.8) even if not stated explicitly. Here it should be clarified that although probability amplitudes (8.5) and (8.6) apply to fermions, equations (8.5)–(8.8) describe the physics of bosons.

Implementations of the Ekert approach extended to bosons have been published by various authors including Naik *et al* (2000), Jennewein *et al* (2000), and Tittel *et al* (2000).

In an adaptation of the Ekert approach for bosons the researchers first select the source to be used and identify this source with one of the corresponding probability amplitudes given in equations (8.5)–(8.8). For instance, for a source that emits quanta pairs as described by equation (8.7) each communicating party, A and B, then proceeds to conduct a set of measurements using states such as (Naik *et al* 2000)

$$(|x\rangle_1 + e^{i\varphi_j^A}|y\rangle_1) \tag{8.9}$$

and

$$(|x\rangle_2 + e^{i\varphi_j^B}|y\rangle_2) \tag{8.10}$$

where the angular quantities φ_j^A and φ_j^B ($j = 1, 2, 3, 4, \ldots$) assume sets of *random* angular settings like $\varphi_1^A = 0$, $\varphi_2^A = \pi/2$, $\varphi_3^A = 3\pi/4$, $\varphi_4^A = \pi/4$ and $\varphi_1^B = \pi/4$, $\varphi_2^B = \pi$, $\varphi_3^B = \pi/2$, $\varphi_4^B = 3\pi/4$.

A and *B* can disclose the settings used for the measurements but not the results of the measurements. Measurements fall into three categories: one portion includes errors and has to be discarded, the second category of measurements is error free and can be used to calculate the corresponding violation of Bell's inequalities. The settings corresponding to this second category are made public. The third category, related to a small fraction of the measurements, correspond to the polarizations that are completely correlated, yielding $(\varphi_j^A + \varphi_j^B) = \pi$, and it is these measurements that are used to provide the *quantum key*. In the Ekert protocol once the quantum key has been established, communications proceed via an open conventional channel (Ekert 1991). In this methodology it is the violation of Bell's inequalities that implies that the measurements are secure.

Bennet *et al* (1993) indicate that the Ekert approach and the Bennet and Brassard approach are equivalent. These authors also argue that *neither* the EPR effect (Einstein *et al* 1935) nor Bell's inequalities (Bell 1964) are 'an essential part' for security certification. Furthermore, it should be indicated that the Pryce–Ward probability amplitude (Pryce and Ward 1947, Ward 1949), indeed an essential component of entanglement based quantum cryptography, was conceived in a total vacuum from philosophical considerations (Duarte 2012, 2014, 2019).

An additional historical note is that the probability amplitude equations given in equations (8.5)–(8.8) became known as *Bell states* even though Bell was absent in their development.

8.3.2 Experiments

The basic apparatus for quantum cryptography, using photons or quanta, is based on the original experimental configuration introduced by Pryce and Ward (1947) and modified by Aspect *et al* (1982) for the testing of Bell's inequalities in the visible spectrum. A relevant generic linear experimental configuration is illustrated in figure 8.4. For instance, a source of entangled photon pairs emits quanta 1 in the $+z$-direction, towards experimenter *A*, and quanta 2 in the $-z$-direction, towards experimenter *B*. In addition to the random emission of photons pairs with polarizations orthogonal to each

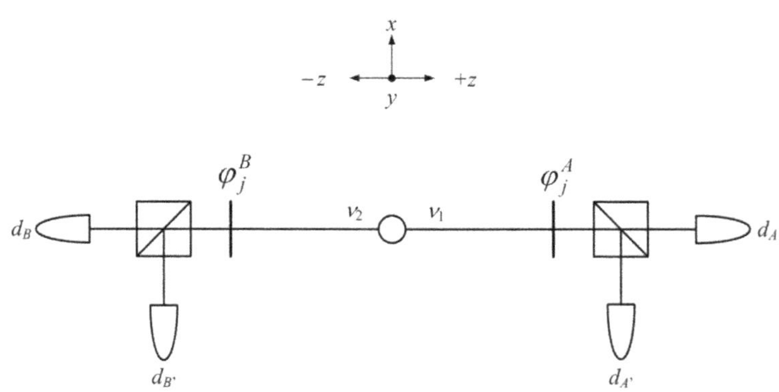

Figure 8.4. Experimental configuration for an idealized quantum key distribution. The. settings of φ_j^A and φ_j^B are randomized (from Duarte 2019).

other, the analyzers are also set randomly at angles φ_j^A and φ_j^B, as outlined in the previous section.

In contemporaneous settings the photon pair sources can either be emitted via an optically-pumped spontaneous parametric down conversion (SPDC) source or quantum dot emitters. SPDC type II sources emit the photon pairs not collinearly in opposite directions but at an angle relative to each other, see figure 8.5. Type I SPDC, using two BBO crystals in series, have been reported to emit nearly collinear photon pairs (Villar *et al* 2020). Quantum dot sources for this application are described by Kuroda *et al* (2013). Also, it should be indicated that researches employ varied methods to achieve the random settings for φ_j^A and φ_j^B.

The quantum entanglement probability applicable to two quanta propagating in opposite directions, while exhibiting orthogonal polarizations, is calculated from the superposition probability amplitude (8.6) and is given by Pryce and Ward (1947)

$$|\psi\rangle|\psi\rangle^* = P(\varphi_1, \varphi_2) = -\cos 2(\varphi_1 - \varphi_2) \tag{8.11}$$

In Bell's inequality the probabilities originally related to hidden variables ultimately are, in an ad hoc manner, replaced by quantum probabilities (Bell 1964). More specifically, for $\varphi_1 = 0$, $\varphi_2 = \pi/3$, $\varphi_1' = \pi/6$, $\varphi_2' = 0$ the probabilities in Bell's expression become quantum superposition probabilities so that (see chapter 10)

$$\Sigma_{|\psi\rangle|\psi\rangle^*} = |P(\varphi_1, \varphi_2) - P(\varphi_1, \varphi_2')| + |P(\varphi_1', \varphi_2') + P(\varphi_1', \varphi_2)| \tag{8.12}$$

with a numerical value of

$$\Sigma_{|\psi\rangle|\psi\rangle^*} = |+0.5 + 1.0| + |-0.5 - 0.5| \tag{8.13}$$

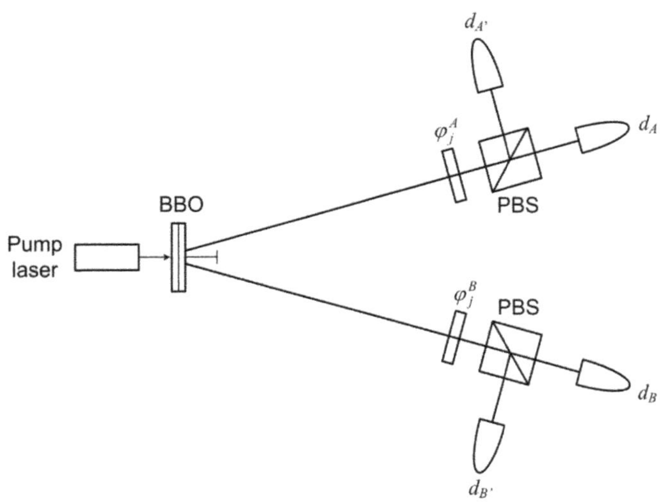

Figure 8.5. Experimental configuration for quantum key distribution utilizing spontaneous. parametric down conversion (SPDC). An appropriate short wavelength pump semiconductor laser is used to excite a type II crystal configuration typically comprised of BBO crystals. A polarizer analyzer capable of providing randomized settings for φ_j^A and φ_j^B is deployed in front of each polarizing beam splitter (PBS). The source of the entangled quanta pairs can be deployed near one of the communicating parties.

$$\Sigma_{|\psi\rangle|\psi\rangle^*} = 2.5 \tag{8.14}$$

and

$$\Sigma_{|\psi\rangle|\psi\rangle^*} \geqslant 2 \tag{8.15}$$

showing that Bell's inequality is violated when using probabilities calculated from the probability amplitudes for quantum entanglement.

Ursin *et al* (2007) conducted an experiment in free-space at a distance of over 144 km.

These authors report clear violation of Bell's inequality with $\Sigma_{|\psi\rangle|\psi\rangle^*} = 2.508 \pm 0.037$ and the implementation of quantum key distribution. Villar *et al* (2020) have used a type I SPDC source to demonstrate polarization quantum entanglement experiments on board a 'nanosatellite' launched from the International Space Station at an altitude of 408 km.

Using the modified Bell approach due to Clauser *et al* (1969) these authors report $\Sigma_{|\psi\rangle|\psi\rangle^*} = 2.63 \pm 0.067$ (Villar *et al* 2020).

8.3.3 All-quantum security protocol

As discussed in chapter 10, hidden variable theories (Bohm 1952a, 1952b, Bohm and Bub 1966) include intrinsic non-quantum features. Based on this key observation the incompatibility of hidden variable theories with quantum mechanics becomes self-evident. In this regard, a simple all-quantum protocol, to verify the security of quantum entanglement communications, can be formulated that is independent of Bell's theorem.

Consider a basic quantum entanglement configuration for, $n = N = 2$, incorporating intruding intra-communication path beam splitters, identified as $\boldsymbol{i}$, as illustrated in figure 8.6. Following the interferometric approach described in chapter 2 (Duarte 2013, 2019) the probability amplitude describing the physics, again assuming identical detectors $d_1 = d_2 = d$, is

$$\langle d|s\rangle = \langle d|p_2\rangle\langle p_2|\boldsymbol{i}_2\rangle\langle\boldsymbol{i}_2|s\rangle + \langle d|p_1\rangle\langle p_1|\boldsymbol{i}_1\rangle\langle\boldsymbol{i}_1|s\rangle \tag{8.16}$$

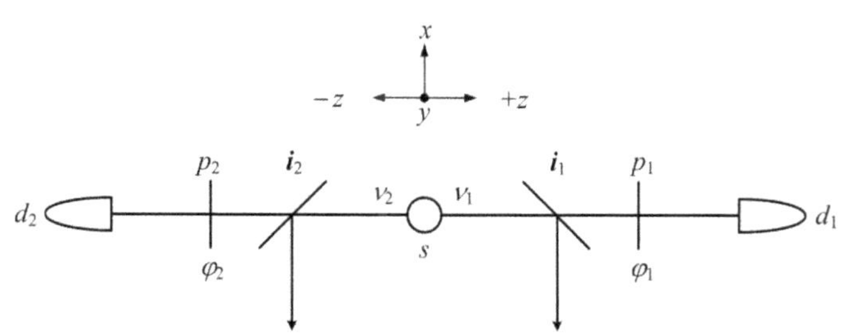

Figure 8.6. Basic quantum entanglement experimental configuration in the optical. domain including intra-path beam splitters for information extraction. The letter $\boldsymbol{i}$ refers to intruder beam splitters. This bold character $\boldsymbol{i}$ should not be confused with the simple i denoting complex numbers.

and abstracting the detectors d leads to

$$|s\rangle = |p_2\rangle\langle p_2|\boldsymbol{i}_2\rangle\langle\boldsymbol{i}_2|s\rangle + |p_1\rangle\langle p_1|\boldsymbol{i}_1\rangle\langle\boldsymbol{i}_1|s\rangle \tag{8.17}$$

Following the approach described in chapter 2 this equation can be re expressed as

$$|s\rangle = |D\rangle_2\langle\boldsymbol{i}_2|s\rangle + |D\rangle_1\langle\boldsymbol{i}_1|s\rangle \tag{8.18}$$

where

$$|D\rangle_1 = |p_1\rangle\langle p_1|\boldsymbol{i}_1\rangle \tag{8.19}$$

and

$$|D\rangle_2 = |p_2\rangle\langle p_2|\boldsymbol{i}_2\rangle \tag{8.20}$$

Setting

$$\langle\boldsymbol{i}_1|s\rangle = e^{-i\theta_1} \tag{8.21}$$

$$\langle\boldsymbol{i}_2|s\rangle = e^{-i\theta_2} \tag{8.22}$$

and defining $|D\rangle_1 = |C\rangle_1$ and $|D\rangle_2 = |C\rangle_2$ the superposition probability amplitude given in equation (8.18) can be written as

$$|s\rangle = |C\rangle_2 e^{-i\theta_2} + |C\rangle_1 e^{-i\theta_1} \tag{8.23}$$

Using the Dirac identities (Dirac 1958) $|C\rangle_1 = |x\rangle_1|y\rangle_2$ and $|C\rangle_2 = |y\rangle_1|x\rangle_2$, and normalizing, equation (8.23) can be expressed as

$$|\psi\rangle_{+i} = 2^{-1/2}(|x\rangle_1|y\rangle_2 e^{-i\theta_1} + |y\rangle_1|x\rangle_2 e^{-i\theta_2}) \tag{8.24}$$

and its linear combination is

$$|\psi\rangle_{-i} = 2^{-1/2}(|x\rangle_1|y\rangle_2 e^{-i\theta_1} - |y\rangle_1|x\rangle_2 e^{-i\theta_2}) \tag{8.25}$$

Immediately it should be noted that for $e^{-i\theta_1} \approx e^{-i\theta_2}$

$$2^{-1/2}(|x\rangle_1|y\rangle_2 e^{-i\theta_1} \pm |y\rangle_1|x\rangle_2 e^{-i\theta_2}) \leqslant 2^{-1/2}(|x\rangle_1|y\rangle_2 \pm y\rangle_1|x\rangle_2) \tag{8.26}$$

and the superposition probability open to intrusion $|\psi\rangle_{+i}|\psi\rangle^*_{+i}$ is related to the intrinsic probability $|\psi\rangle_+|\psi\rangle^*_+$ by

$$|\psi\rangle_{+i}|\psi\rangle^*_{+i} \leqslant |\psi\rangle_+|\psi\rangle^*_+ \tag{8.27}$$

$$|\psi\rangle_{-i}|\psi\rangle^*_{-i} \leqslant |\psi\rangle_-|\psi\rangle^*_- \tag{8.28}$$

For the optimum case in the absence of intrusion

$$e^{-i\theta_1} \approx e^{-i\theta_2} \approx 1 \tag{8.29}$$

and the superposition probabilities become

$$|\psi\rangle_{\pm i}|\psi\rangle^*_{\pm i} \approx |\psi\rangle_\pm|\psi\rangle^*_\pm \tag{8.30}$$

The Pryce–Ward superposition probability, calculated from $|\psi\rangle_{-}|\psi\rangle_{-}^{*}$, is given by (Pryce and Ward 1947, Duarte 2019)

$$|\psi\rangle_{-}|\psi\rangle_{-}^{*} = P(\varphi_1, \varphi_2) = -\cos 2(\varphi_1 - \varphi_2) \tag{8.31}$$

and for the four possible angular combinations (φ_1, φ_2), (φ_1, φ_2'), (φ_1', φ_2'), (φ_1', φ_2) the quantum superposition probabilities are

$$P(\varphi_1, \varphi_2) = -\cos 2(\varphi_1 - \varphi_2) \tag{8.32}$$

$$P(\varphi_1, \varphi_2') = -\cos 2(\varphi_1 - \varphi_2') \tag{8.33}$$

$$P(\varphi_1', \varphi_2') = -\cos 2(\varphi_1' - \varphi_2') \tag{8.34}$$

$$P(\varphi_1', \varphi_2) = -\cos 2(\varphi_1' - \varphi_2) \tag{8.35}$$

The overall quantum probability of the experiment is given by the straightforward addition of the absolute value of *each of the four* quantum superposition probabilities

$$\Sigma_{|\psi\rangle_{-}|\psi\rangle_{-}^{*}} = |P(\varphi_1, \varphi_2)| + |P(\varphi_1, \varphi_2')| + |P(\varphi_1', \varphi_2')| + |P(\varphi_1', \varphi_2)| \tag{8.36}$$

which is clearly different to the probability equation given by Bell (1964) and included previously as equation (8.12).

As an example consider the following angular settings: $\varphi_1 = \pi/20$, $\varphi_2 = 9\pi/20$, $\varphi_1' = \pi/3$, $\varphi_2' = \pi/6$. This leads directly to

$$\Sigma_{|\psi\rangle_{-}|\psi\rangle_{-}^{*}} = |0.8090| + |-0.7431| + |-0.5000| + |-0.7431| = 2.7952 \tag{8.37}$$

All-quantum measurement protocol:

1. Perform a series of measurements, in free space, whilst actively and continuously monitoring that no third party intruders are present. Thus arrive at an experimentally determined *intrinsic quantum superposition probability* (IQSP)

$$\Sigma_{|\psi\rangle-|\psi\rangle_{-}^{*}} \pm \Delta\Sigma_{|\psi\rangle-|\psi\rangle_{-}^{*}} \tag{8.38}$$

2. The previous step also yields a set of completely correlated measurements that can be used as the key.
3. According to inequality (8.26) beam-splitter intrusion, or optical eavesdropping, yields a measured probability $\Sigma_{|\psi\rangle_{-i}|\psi\rangle_{-i}^{*}} \pm \Delta\Sigma_{|\psi\rangle_{-i}|\psi\rangle_{-i}^{*}}$ which is less than the lower limit of the IQSP, or

$$\Sigma_{|\psi\rangle_{-i}|\psi\rangle_{-i}^{*}} \pm \Delta\Sigma_{|\psi\rangle_{-i}|\psi\rangle_{-i}^{*}} \leqslant \Sigma_{|\psi\rangle_{-}|\psi\rangle_{-}^{*}} - \Delta_{\Sigma|\psi\rangle_{-}|\psi\rangle_{-}^{*}} \tag{8.39}$$

4. Once the IQSP has been determined, only polarization data giving rise to measured probabilities within the $\Sigma_{|\psi\rangle_{-}|\psi\rangle_{-}^{*}} \pm \Delta\Sigma_{|\psi\rangle_{-}|\psi\rangle_{-}^{*}}$ range are to be considered as secure. Polarization data leading to probabilities outside the intrinsic measured range (8.38), is to be discarded.

It should be re-emphasized that the crucial IQSP, or $\Sigma_{|\psi\rangle_{-}|\psi\rangle_{-}^{*}} \pm \Delta\Sigma_{|\psi\rangle_{-}|\psi\rangle_{-}^{*}}$, is an experimental quantity to be determined by measurements exclusively, which is in total agreement with Born's quantum experimental doctrine (Born 1949). This simple measurement-based all-quantum methodology is a more transparent and more rigorous approach than the use of Bell's theorem which ultimately requires only that $\Sigma_{|\psi\rangle_{-}|\psi\rangle_{-}^{*}} \geqslant 2$ in order to 'certify' security in the Ekert protocol.

8.4 Quantum entanglement applications to teleportation

Teleportation of a quantum state, from one position in space to another position in space, is based on the physics of the Pryce–Ward probability amplitude

$$|\psi\rangle = 2^{-1/2}(|x\rangle_1|y\rangle_2 \pm |y\rangle_1|x\rangle_2) \tag{8.40}$$

More specifically, *quantum teleportation*, as introduced by Bennett *et al* (1993), consists in the disassembling of one quantum state, a polarization state at the emitter's site, and the subsequent reassembling, of that very quantum state at the receiver's site. This exercise is performed at the probability amplitude level. It is the disassembling and reassembling of a mathematical expression.

Here, the polarization states $|x\rangle$ and $|y\rangle$ are represented by $|1\rangle$ and $|0\rangle$, and their corresponding vectors (Fowles 1968)

$$|x\rangle = |1\rangle = \begin{pmatrix} 1 \\ 0 \end{pmatrix} \tag{8.41}$$

and

$$|y\rangle = |0\rangle = \begin{pmatrix} 0 \\ 1 \end{pmatrix} \tag{8.42}$$

For reviews on the subject, including technology trends, the reader may refer to Pirandola *et al* (2015).

8.4.1 The methodology of teleportation

Here, the mechanics of disassembling a quantum state at the emitter's site and the subsequent reassembling of the same quantum state at the receiver's site is described.

This exposition uses the style and notation of a previous review on the subject (Duarte 2014, 2019). As necessary background, the 2×2 identity matrix I and Pauli's matrices are restated (Feynman *et al* 1965)

$$I = \begin{pmatrix} 1 & 0 \\ 0 & 1 \end{pmatrix} \tag{8.43}$$

$$\sigma_x = \begin{pmatrix} 0 & 1 \\ 1 & 0 \end{pmatrix} \tag{8.44}$$

$$\sigma_y = \begin{pmatrix} 0 & -i \\ i & 0 \end{pmatrix} \tag{8.45}$$

$$\sigma_z = \begin{pmatrix} 1 & 0 \\ 0 & -1 \end{pmatrix} \tag{8.46}$$

Using vector notation, the Pryce–Ward probability amplitudes can be expressed as (Duarte 2019)

$$|s\rangle_+ = 2^{-1/2}(|1\rangle_1|0\rangle_2 + |0\rangle_1|1\rangle_2) \tag{8.47}$$

$$|s\rangle_- = 2^{-1/2}(|1\rangle_1|0\rangle_2 - |0\rangle_1|1\rangle_2) \tag{8.48}$$

$$|r\rangle_+ = 2^{-1/2}(|1\rangle_1|1\rangle_2 + |0\rangle_1|0\rangle_2) \tag{8.49}$$

$$|r\rangle_- = 2^{-1/2}(|1\rangle_1|1\rangle_2 - |0\rangle_1|0\rangle_2) \tag{8.50}$$

Adding and subtracting these probability amplitudes yield the following combined states

$$2^{-1/2}(|s\rangle_+ + |s\rangle_-) = |1\rangle_1|0\rangle_2 \tag{8.51}$$

$$2^{-1/2}(|s\rangle_+ - |s\rangle_-) = |0\rangle_1|1\rangle_2 \tag{8.52}$$

$$2^{-1/2}(|r\rangle_+ + |r\rangle_-) = |1\rangle_1|1\rangle_2 \tag{8.53}$$

$$2^{-1/2}(|r\rangle_+ - |r\rangle_-) = |0\rangle_1|0\rangle_2 \tag{8.54}$$

The mechanics of teleporting one quantum state from the transmitter's site to the receiver's site can be illustrated using the scheme suggested by Kim *et al* (2001).

Assume that the following arbitrary quantum state is to be teleported

$$|\phi\rangle_1 = (\alpha|1\rangle_1 + \beta\,|0\rangle_1) \tag{8.55}$$

In explicit vector notation this state can also be expressed as

$$|\phi\rangle_1 = \left(\begin{pmatrix} \alpha \\ 0 \end{pmatrix}_1 + \begin{pmatrix} 0 \\ \beta \end{pmatrix}_1\right) = \begin{pmatrix} \alpha \\ \beta \end{pmatrix}_1 \tag{8.56}$$

Here, the α and β factors are related by $|\alpha|^2 + |\beta|^2 = 1$. This probability amplitude, $|\phi\rangle_1 = (\alpha|1\rangle_1 + \beta\,|0\rangle_1)$ is the state to be disassembled at the emitter's site and it is also the state to be reassembled at the receiver's site. This is what constitutes the action of *teleportation.*

The emitter can generate an entangled state of the form of any of the probability amplitudes given in equations (8.47)–(8.50). Assume that the entangled state selected to perform the teleportation operation is that given in equation (8.47). Replacing $|s\rangle_+$ by $|s\rangle_{23+}$

$$|s\rangle_{23+} = 2^{-1/2}(|1\rangle_2|0\rangle_3 + |0\rangle_2|1\rangle_3) \tag{8.57}$$

A three-quanta state $|\varphi\rangle_{123}$ can be created by the emitter via the multiplication of $|\phi\rangle_1$ and $|s\rangle_{23+}$

$$|\varphi\rangle_{123} = 2^{-1/2}(\alpha\ |1\rangle_1|1\rangle_2|0\rangle_3 + \alpha\ |1\rangle_1|0\rangle_2|1\rangle_3 + \beta\ |0\rangle_1|1\rangle_2|0\rangle_3 + \beta\ |0\rangle_1|0\rangle_2|1\rangle_3) \quad (8.58)$$

This operation makes possible the projection of photon 3 on the input state given in equations (8.54) and (8.55). Using equations (8.51)–(8.54) while noting that the quantum entanglement states: $|s\rangle_+ = |s\rangle_{12+}$, $|s\rangle_- = |s\rangle_{12-}$, $|r\rangle_+ = |r\rangle_{12+}$, $|r\rangle_- = |r\rangle_{12-}$, then the three quanta state $|\varphi\rangle_{123}$, that is, equation (8.58) can be rewritten as

$$|\varphi\rangle_{123} = 2^{-1}(|s\rangle_{12+}|\phi\rangle_{3+} + |s\rangle_{12-}|\phi\rangle_{3-} + |r\rangle_{12+}|\vartheta\rangle_{3+} + |r\rangle_{12-}|\vartheta\rangle_{3-}) \quad (8.59)$$

indicating that each of the four component states has an equal 1/4 chance of existing.

In equation (8.55) photon 3 is projected into the following states

$$|\phi\rangle_{3+} = (\alpha|1\rangle_3 + \beta\ |0\rangle_3) \quad (8.60)$$

$$|\phi\rangle_{3-} = (\alpha|1\rangle_3 - \beta\ |0\rangle_3) \quad (8.61)$$

$$|\vartheta\rangle_{3+} = (\alpha|0\rangle_3 + \beta\ |1\rangle_3) \quad (8.62)$$

$$|\vartheta\rangle_{3-} = (\alpha|0\rangle_3 - \beta\ |1\rangle_3) \quad (8.63)$$

or

$$|\phi\rangle_{3+} = \begin{pmatrix}\alpha\\ \beta\end{pmatrix}_3 \quad (8.64)$$

$$|\phi\rangle_{3-} = \begin{pmatrix}\alpha\\ -\beta\end{pmatrix}_3 \quad (8.65)$$

$$|\vartheta\rangle_{3+} = \begin{pmatrix}\beta\\ \alpha\end{pmatrix}_3 \quad (8.66)$$

$$|\vartheta\rangle_{3-} = \begin{pmatrix}-\beta\\ \alpha\end{pmatrix}_3 \quad (8.67)$$

Following the multiplication of $|\phi\rangle_1$ and $|s\rangle_{23+}$, the original state to be teleported has been disassembled and now photon 3 is projected on the original input state.

In this methodology if the emitter measures the probability corresponding to $|s\rangle_{12+}$, the qbit communicated to the receiver is $|\phi\rangle_{3+}$. In this case, the only unitary transformation needed is the identity matrix I. If, on the other hand, $|s\rangle_{12-}$ is communicated then the receiver must apply the appropriate unitary transformation on $|\phi\rangle_{3-}$ to recover the original quantum state. In each case the necessary unitary transformation is communicated classically by the emitter to the receiver (Kim *et al* 2001).

This mechanics can be summarized as $|s\rangle_{12+} \rightarrow |\phi\rangle_{3+}$, where $\rightarrow$ stands for 'communicate to receiver': The mechanics for the four cases $|s\rangle_{12+} \rightarrow |\phi\rangle_{3+}$, $|s\rangle_{12-} \rightarrow |\phi\rangle_{3-}$, $|r\rangle_{12+} \rightarrow |\vartheta\rangle_{3+}$, and $|r\rangle_{12-} \rightarrow |\vartheta\rangle_{3-}$ is summarized next.

$|s\rangle_{12+} \rightarrow |\phi\rangle_{3+}$:

$$I\,|\phi\rangle_{3+} = (\alpha|1\rangle_3 + \beta\,|0\rangle_3) \tag{8.68}$$

$$I\,|\phi\rangle_{3+} = \begin{pmatrix} 1 & 0 \\ 0 & 1 \end{pmatrix}\begin{pmatrix} \alpha \\ \beta \end{pmatrix}_3 = \begin{pmatrix} \alpha \\ \beta \end{pmatrix}_3 \tag{8.69}$$

$$I\,|\phi\rangle_{3+} = |\phi\rangle_{3+} \tag{8.70}$$

$|s\rangle_{12-} \rightarrow |\phi\rangle_{3-}$:

$$\sigma_z\,|\phi\rangle_{3-} = \sigma_z(\alpha|1\rangle_3 - \beta\,|0\rangle_3) \tag{8.71}$$

$$\begin{pmatrix} 1 & 0 \\ 0 & -1 \end{pmatrix}\begin{pmatrix} \alpha \\ -\beta \end{pmatrix}_3 = \begin{pmatrix} \alpha \\ \beta \end{pmatrix}_3 \tag{8.72}$$

$$\sigma_z\,|\phi\rangle_{3-} = |\phi\rangle_{3+} \tag{8.73}$$

$|r\rangle_{12+} \rightarrow |\vartheta\rangle_{3+}$:

$$\sigma_x\,|\vartheta\rangle_{3+} = \sigma_x(\alpha|0\rangle_3 + \beta\,|1\rangle_3) \tag{8.74}$$

$$\begin{pmatrix} 0 & 1 \\ 1 & 0 \end{pmatrix}\begin{pmatrix} \beta \\ \alpha \end{pmatrix}_3 = \begin{pmatrix} \alpha \\ \beta \end{pmatrix}_3 \tag{8.75}$$

$$\sigma_x\,|\vartheta\rangle_{3+} = |\phi\rangle_{3+} \tag{8.76}$$

$|r\rangle_{12-} \rightarrow |\vartheta\rangle_{3-}$:

$$i\sigma_y\,|\vartheta\rangle_{3-} = i\sigma_y(\alpha|0\rangle_3 - \beta\,|1\rangle_3) \tag{8.77}$$

$$i\begin{pmatrix} 0 & -i \\ i & 0 \end{pmatrix}\begin{pmatrix} -\beta \\ \alpha \end{pmatrix}_3 = \begin{pmatrix} \alpha \\ \beta \end{pmatrix}_3 \tag{8.78}$$

$$i\sigma_y\,|\vartheta\rangle_{3-} = |\phi\rangle_{3+} \tag{8.79}$$

Again, the states $|\phi\rangle_{3+}$, $|\phi\rangle_{3-}$, $|\vartheta\rangle_{3+}$, and $|\vartheta\rangle_{3-}$ are transmitted quantum mechanically and the unitary transformations to recover $|\phi\rangle_{3+}$ are $I\,|\phi\rangle_{3+}$, $\sigma_z\,|\phi\rangle_{3-}$, $\sigma_x\,|\vartheta\rangle_{3+}$, and $i\sigma_y\,|\vartheta\rangle_{3-}$. In each case the necessary unitary operators I, σ_z, σ_x, and $i\sigma_y$ are transmitted via a classical channel. A simplified schematic of the quantum state teleportation scheme is outlined in figure 8.7.

Recapping: the emitter starts with photon 1 in the state $|\phi\rangle_1 = (\alpha|1\rangle_1 + \beta\,|0\rangle_1)$ and proceeds to disassemble this state via a transformation which multiplies $|\phi\rangle_1$ with $|s\rangle_{23+}$. Next, a series of additional states are created, such as $|\phi\rangle_{3+}$, $|\phi\rangle_{3-}$, $|\vartheta\rangle_{3+}$, and $|\vartheta\rangle_{3-}$. These states are transmitted quantum mechanically to the receiver. In turn,

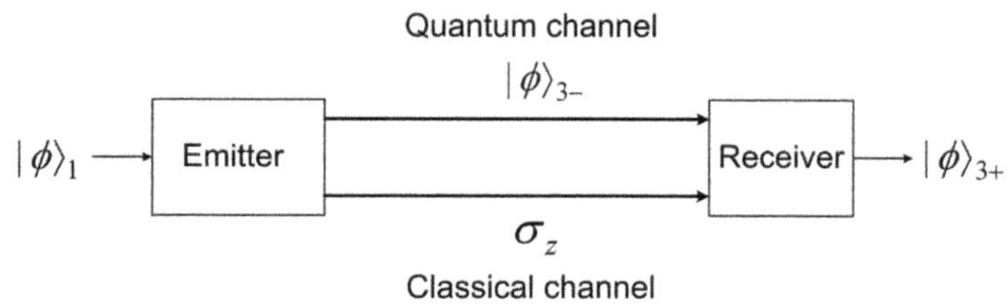

Figure 8.7. Simplified experimental configuration for quantum teleportation: The quantum state $|\phi\rangle_1$ (the original state) and $|\phi\rangle_{3+}$ (the teleported state) are identical via the use of the appropriate unitary transformation, i. e. $|\phi\rangle_1 \equiv |\phi\rangle_{3+}$. The transmission channel for the $|\phi\rangle_{3-}$ state is a quantum channel while σ_z is sent to the receiver via a classical channel (from Duarte 2019).

using any of these states the receiver can recreate the original state $|\phi\rangle_1$ now represented by photon 3 using the appropriate unitary transformation.

8.4.2 Technology

The unitary transformations are performed using electro-optical modulators (EOMs) (Ma *et al* 2012). Furthermore, quantum teleportation experiments are rather sophisticated and depend on numerous technologies including lasers, photon pair sources, half wave plates, quarter wave plates, polarizing beam splitters, fiber beam splitters, prisms, mirrors, telescopes, and single-photon detectors. As an example, the experiment performed by Ma *et al* (2012), over a distance of 143 km, included more than 60 optics and electro-optics components. These researchers report teleportation fidelities in the 0.76 – 0.80 range. Yin *et al* (2012) report on experiments over a distance of 100 km. The use of quantum dot sources generating $|\psi\rangle^+ = 2^{-1/2}(|x\rangle_1|x\rangle_2 + |y\rangle_1|y\rangle_2)$ states has been reported by Rota *et al* (2020).

8.5 Quantum computing

Some of the first concepts on quantum computing were introduced by Feynman (1985, 1986). Feynman's approach to quantum computing was his own and thus it did not follow the now established philosophical path. The concept of a quantum bit was abbreviated into qubit is by Schumacher (1995). Qubit is also known as *qbit.*

Subsequently, Steane (1998) dealt directly with EPR–Bell notions. Extensive reviews on quantum computing are given by Williams and Clearwater (1998) and Nielsen and Chuang (2000).

Multiple entangled states, such as $n = N = 2^1, 2^2, 2^3, 2^4, \ldots, 2^r$, are considered in chapter 2. Studies with potential applications to large scale processors also include understanding the interaction between distant multiple qbits (Vadimov *et al* 2020).

One of the most promising advantages of quantum computing relates to the enormous time savings to be expected in the computation of specific physics. For instance, computing the interferometric propagation of a single photon, or an ensemble of indistinguishable photons, via an array of $N = 1500$ slits using an optimally programmed universal computer takes $t_u = 3112.20$ s. Using the corresponding interferometric computer, which is analogous to a quantum computer, takes approximately $t_q \approx 30 \times 10^{-3}$ s (Duarte 2003, 2014). This represents an improvement in computational time better than $t_u/t_q \approx 10^5$. Certainly, most of the time taken in the $t_q \approx 30 \times 10^{-3}$ s is due to electronic processing that can be improved ever further rather significantly.

In this section aspects of quantum computing directly related to quantum entanglement are examined. The presentation is rather succinct. For matrix derivational aspects the reader should refer to chapter 7 and Duarte (2019).

8.5.1 Pauli gates

There are three Pauli gates, that are based on the mathematical properties of the Pauli matrices (σ_x, σ_y, σ_z).

$$\sigma_x = \begin{pmatrix} 0 & 1 \\ 1 & 0 \end{pmatrix}$$

$$\sigma_y = \begin{pmatrix} 0 & -i \\ i & 0 \end{pmatrix}$$

$$\sigma_z = \begin{pmatrix} 1 & 0 \\ 0 & -1 \end{pmatrix}$$

The Pauli-X gate is equivalent to the NOT gate and it induces a polarization rotation of $\pi/2$ radians. In other words, it is a polarization rotator which converts $|x\rangle$ to $|y\rangle$, or $|y\rangle$ to $|x\rangle$. Specifically,

$$\begin{pmatrix} 0 & 1 \\ 1 & 0 \end{pmatrix}\begin{pmatrix} 1 \\ 0 \end{pmatrix} = \begin{pmatrix} 0 \\ 1 \end{pmatrix} \tag{8.80}$$

$$\begin{pmatrix} 0 & 1 \\ 1 & 0 \end{pmatrix}\begin{pmatrix} 0 \\ 1 \end{pmatrix} = \begin{pmatrix} 1 \\ 0 \end{pmatrix} \tag{8.81}$$

which in equivalent notation are

$$\sigma_x|1\rangle \to |0\rangle \tag{8.82}$$

$$\sigma_x|0\rangle \to |1\rangle \tag{8.83}$$

The Pauli-Y gate is described by the transformations

$$\begin{pmatrix} 0 & -i \\ i & 0 \end{pmatrix}\begin{pmatrix} 1 \\ 0 \end{pmatrix} = \begin{pmatrix} 0 \\ i \end{pmatrix} \tag{8.84}$$

$$\begin{pmatrix} 0 & -i \\ i & 0 \end{pmatrix}\begin{pmatrix} 0 \\ 1 \end{pmatrix} = \begin{pmatrix} -i \\ 0 \end{pmatrix} \tag{8.85}$$

which in equivalent notation are

$$\sigma_y|1\rangle \to i|0\rangle \tag{8.86}$$

$$\sigma_y|0\rangle \rightarrow -i|1\rangle \tag{8.87}$$

The Pauli-Z gate is described by the transformations

$$\begin{pmatrix} 1 & 0 \\ 0 & -1 \end{pmatrix}\begin{pmatrix} 1 \\ 0 \end{pmatrix} = \begin{pmatrix} 1 \\ 0 \end{pmatrix} \tag{8.88}$$

$$\begin{pmatrix} 1 & 0 \\ 0 & -1 \end{pmatrix}\begin{pmatrix} 0 \\ 1 \end{pmatrix} = \begin{pmatrix} 0 \\ -1 \end{pmatrix} \tag{8.89}$$

which in equivalent notation are

$$\sigma_z|1\rangle \rightarrow |1\rangle \tag{8.90}$$

$$\sigma_z|0\rangle \rightarrow -|0\rangle \tag{8.91}$$

8.5.2 The Pauli-X gate and quantum entanglement

For the sake of transparency, in this section the quantum entanglement probability amplitudes are again stated explicitly as

$$|\psi\rangle_+ = 2^{-1/2}(|1\rangle_1|0\rangle_2 + |0\rangle_1|1\rangle_2) \tag{8.92}$$

$$|\psi\rangle_- = 2^{-1/2}(|1\rangle_1|0\rangle_2 - |0\rangle_1|1\rangle_2) \tag{8.93}$$

$$|\psi\rangle^+ = 2^{-1/2}(|1\rangle_1|1\rangle_2 + |0\rangle_1|0\rangle_2) \tag{8.94}$$

$$|\psi\rangle^- = 2^{-1/2}(|1\rangle_1|1\rangle_2 - |0\rangle_1|0\rangle_2) \tag{8.95}$$

Next, the Pauli-X gate, which is equivalent to a $\pi/2$ polarization rotator, such as a half-wave plate, a double Fresnel rhomb, or a multiple-prism broadband rotator (Duarte 1989, 2019), with a matrix represented by

$$\sigma_x = R = \begin{pmatrix} 0 & 1 \\ 1 & 0 \end{pmatrix} \tag{8.96}$$

perform the following transformations when operating on quantum entanglement states

$$\sigma_x|\psi\rangle^+ = R|\psi\rangle^+ = 2^{-1/2}\sigma_x \tag{8.97}$$

$$\sigma_x|\psi\rangle^- = R|\psi\rangle^- = -2^{-1/2}i\sigma_y \tag{8.98}$$

$$\sigma_x|\psi\rangle_+ = R\,|\psi\rangle_+ = 2^{-1/2}I \tag{8.99}$$

$$\sigma_x|\psi\rangle_- = R|\psi\rangle_- = -2^{-1/2}\sigma_z \tag{8.100}$$

which are equivalent to

$$R|\psi\rangle^+ = |\psi\rangle_+ \tag{8.101}$$

$$R|\psi\rangle^{-} = -|\psi\rangle_{-} \tag{8.102}$$

$$R\,|\psi\rangle_{+} = |\psi\rangle^{+} \tag{8.103}$$

$$R|\psi\rangle_{-} = -|\psi\rangle^{-} \tag{8.104}$$

8.5.3 The Hadamard gate and quantum entanglement

The Hadamard gate discussed in chapter 7, and whose logic symbol is depicted in figure 8.8, is represented by the matrix

$$H_G = 2^{-1/2}\begin{pmatrix} 1 & 1 \\ 1 & -1 \end{pmatrix} \tag{8.105}$$

Using the matrix notation introduced in chapter 7 it can be shown that the Hadamard gate H_G operating on $|\psi\rangle_{+}$, $|\psi\rangle_{-}$, $|\psi\rangle^{+}$, and $|\psi\rangle^{-}$ yields the following mathematical identities (Duarte *et al* 2020)

$$H_G|\psi\rangle_{-} = 2^{-1/2}(I + i\sigma_y) \tag{8.106}$$

$$H_G|\psi\rangle_{-} = 2^{-1}(\sigma_x - \sigma_z) \tag{8.107}$$

$$H_G|\psi\rangle^{+} = 2^{-1}(\sigma_x + \sigma_z) \tag{8.108}$$

$$H_G|\psi\rangle^{-} = 2^{-1}(I - i\sigma_y) \tag{8.109}$$

These mathematical identities can also be expressed as (Duarte *et al* 2020)

$$H_G|\psi\rangle_{+} = 2^{-1/2}(|\psi\rangle^{+} + |\psi\rangle_{-}) \tag{8.110}$$

$$H_G|\psi\rangle_{-} = 2^{-1/2}(|\psi\rangle_{+} - |\psi\rangle^{-}) \tag{8.111}$$

$$H_G|\psi\rangle^{+} = 2^{-1/2}(|\psi\rangle_{+} + |\psi\rangle^{-}) \tag{8.112}$$

$$H_G|\psi\rangle^{-} = 2^{-1/2}(|\psi\rangle^{+} - |\psi\rangle_{-}) \tag{8.113}$$

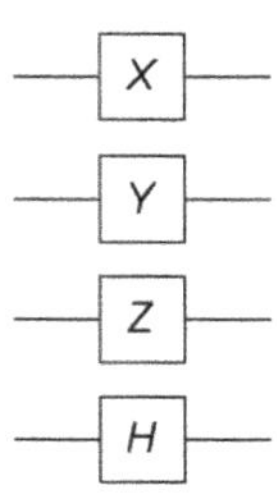

Figure 8.8. Generic quantum computing symbols for the Pauli-X, Pauli-Y, Pauli-Z, and Hadamard gates (from Duarte 2019).

Applying the matrix representation of the entanglement states introduced in chapter 7 it can also be shown that

$$H_G = (|\psi\rangle_+ + |\psi\rangle^-) \tag{8.114}$$

$$H_G = 2^{-1/2}(\sigma_x + \sigma_z) \tag{8.115}$$

The mathematical operations symbolized in equations (8.101)–(8.104) and (8.110)–(8.113) are intrinsic quantum mathematical operations.

8.5.4 Technology

It has been known for a while that quantum-optics based computing techniques can perform interferometric calculations in the millisecond regime that take hours to be performed in optimally-programmed conventional universal computers (Duarte 1993, 2003). Certainly, the promise of quantum computing is beyond enormous.

Central to the technology of optical quantum computing is the capability to handle large number of qbits in a low noise environment. The intent is to generate large numbers of qbits, in the $10^3 \leqslant N_{qbit} \leqslant 10^6$ range, routinely with high fidelity (Duarte 2019). Once this task is completed most likely the effort will turn to miniaturization.

The technology of quantum computing is evolving so fast that that the only chance not to become immediately obsolete is to limit remarks as much as possible. As this book goes to print non-optical big-tech approaches to quantum computing have garnered a lot of media attention. However, optically based quantum computing should not be dismissed in its ability to handle a large number of qbits in the $10^3 \leqslant N_{\text{qbit}} \leqslant 10^6$ range, or even more.

8.6 Quantum entanglement applications to metrology

From a broad perspective, quantum metrology is a field that applies quantum physics to metrology sub disciplines such as interferometry and spectrometry (Duarte and Paine 1989, Duarte 1993). A subfield within that broader context is quantum entanglement metrology that, or instance, applies quantum entanglement states of light to interferometry.

This approach is motivated by the observation that the noise-limit allowed by the uncertainty principle in a quantum entanglement interferometer, such as a Mach–Zehnder interferometer, scales as N^{-1} while the equivalent noise limit generated with standard illumination scales as $N^{-1/2}$, where N is the number of quanta (Lee *et al* 2002). This means that the quantum entanglement approach offers an improvement of $N^{-1/2}$ over the standard approach. Noise reduction is the key motivating factor in quantum entanglement interferometry.

A widely used notational tool in this field is the so-called *NOON* states which apply to ensembles of bosons, or quanta, and are defined as (Lee *et al* 2002)

$$|N, 0\rangle = |0\rangle_1|0\rangle_2|0\rangle_3\ldots|0\rangle_N \tag{8.116}$$

$$|0, N\rangle = |1\rangle_1|1\rangle_2|1\rangle_3\ldots|1\rangle_N \tag{8.117}$$

The use of *NOON* states for quantum metrology applications in biology is discussed by Taylor and Bowen (2016).

For two-photon interference the superposition for the NOON state is written as (Ou 2017)

$$|NOON\rangle = 2^{-1/2}(|2\rangle_x|0\rangle_y + |0\rangle_x|2\rangle_y) \tag{8.118}$$

Switching from polarization to denoting a beam splitter with ports 1 and 2, the notation used by some authors is (Ou 2017)

$$|\psi\rangle = 2^{-1/2}(|2\rangle_1|0\rangle_2 \pm |0\rangle_1|2\rangle_2) \tag{8.119}$$

If a phase shift is introduced in one of the arms of a Mach–Zehnder interferometer the situation is described as (Taylor and Bowen 2016)

$$|\psi\rangle = 2^{-1/2}(e^{i2\phi}|2\rangle_1|0\rangle_2 \pm |0\rangle_1|2\rangle_2) \tag{8.120}$$

It should be indicated that some authors note that if technical noise dominates it is difficult to achieve improvements via quantum entanglement and even if the technical noise is below the quantum limit improvements are 'constrained' (Taylor and Bowen 2016).

8.7 Overview

In addition to cryptographic communications, teleportation of quantum sates, quantum computing, and quantum metrology, the physics of $(|x_1, y_2\rangle - |y_1, x_2\rangle)$ has also found applications, via Michelson interferometry, in proposals to improve the sensitivity of gravitational wave detectors such as LIGO (Schnabel *et al* 2010).

Also, in the area of astronomy, quantum entanglement experiments using high-red-shifted quasar sources have been performed by Rauch *et al* (2018). Quantum entanglement has also given rise to highly speculative physics, which nevertheless is also intriguing and fascinating. In 1957 John Wheeler coined a term that is widely used in science fiction: *wormhole* (Misner and Wheeler 1957). This concept is based on one of the solutions to the equations of general relativity (Einstein and Rosen 1935, Misner *et al* 2017) which is also known as the Einstein–Rosen (ER) bridge, or ER bridge. More recently, this concept has been conceptually linked to quantum entanglement (Maldacena and Susskind 2013). However there is no mathematical proof that the ER bridge is equivalent to quantum entanglement, or that ER = EPR. In this regard, it is clearly evident that the physics of general relativity and the physics of quantum entanglement remain divergent.

Also in the area of astronomy, theoreticians have proposed to apply quantum entanglement principles to measure quantum gravitational effects via the Bose–Marletto–Vedral effect (Tamburini and Licata 2020).

Researchers working in the field of quantum biology (Davies 2009, McFadden and Al-Khalili 2014) which appears to have been inspired by Schrödinger's book *What is Life?* (Schrödinger 1944) think that 'non-trivial quantum phenomena, such as superposition, entanglement, and tunneling might be important to life' (Davies 2020).

Quantum interference should be added to this set of phenomena, albeit it is already linked to superposition and entanglement.

The matrix equations derived from quantum entanglement have also found applications in the development of techniques for quantum approximate optimization algorithms of interest in wireless engineering (Choi *et al* 2020).

If consolidated via concrete technological developments, at the engineering level, the physics of quantum entanglement could spur gigantic, and yet unimagined, advances in present day technology.

Problems

8.1 Using $\varphi_1 = 0$, $\varphi_2 = \pi/3$, $\varphi_1{}' = \pi/6$, $\varphi_2{}' = 0$ verify that the numerical result given in equation (8.14) is correct.

8.2 Verify the equation for $|\vartheta\rangle_{123}$.

8.3 Using the quantum entanglement probability amplitudes given in equations (8.47)–(8.50) derive equations (8.51)–(8.54).

8.4 Using equation (8.58) and equations (8.51)–(8.54) to derive equation (8.59).

8.5 If the original state is given by $|\phi\rangle_1 = (\alpha|0\rangle_1 + \beta\,|1\rangle_1)$ derive equations for $|\phi\rangle_{3+}$, $|\phi\rangle_{3-}$, $|\vartheta\rangle_{3+}$, and $|\vartheta\rangle_{3-}$.

8.6 Using the new found states for $|\phi\rangle_{3+}$, $|\phi\rangle_{3-}$, $|\vartheta\rangle_{3+}$, and $|\vartheta\rangle_{3-}$ derive the equations for $I\,|\phi\rangle_{3+}$, $\sigma_z\,|\phi\rangle_{3-}$, $\sigma_x\,|\vartheta\rangle_{3+}$, and $i\sigma_y\,|\vartheta\rangle_{3-}$.

8.7 Show that equations (8.97)–(8.100) are completely equivalent to equations (8.101)–(8.104).

8.8 Show that equations (8.106)–(8.109) are completely equivalent to equations (8.110)–(8.113).

References

Aspect A, Grangier P and Roger G 1982 Experimental test of Bell's inequality using time-varying analyzers *Phys. Rev. Lett.* **49** 1804–7

Bell J S 1964 On the Einstein–Podolsky–Rosen paradox *Physics (N.Y.)* **1** 195–200

Bennett C H 1992 Quantum cryptography using any two nonorthogonal states *Phys. Rev. Lett.* **68** 3121–24

Bennett C H, Bessette F, Brassard G, Salvail L and Smolin J 1992 Experimental quantum cryptography *J. Cryptol.* **5** 3–28

Bennett C H, Brassard G, Crépeau C, Jozsa R, Peres A and Wootters W K 1993 Teleporting an unknown quantum state via dual classical and Einstein–Podolsky–Rosen channels *Phys. Rev. Lett.* **70** 1895–99

Bennett C H and Brassard G 1984 Quantum cryptography: public key distribution and coin tossing *Proc. IEEE Int. Conf. on Computers Systems and Signal Processing (Bangalore, India)*

Bennett C H and Brassard G 2014 Quantum cryptography: public key distribution and coin tossing *Theo. Comp. Sci.* **560** 7–11

Bennett C H, Brassard G and Mermin N D 1992b Quantum cryptography without Bells theorem *Phys. Rev. Lett.* **68** 557–59

Bohm D 1951 *Quantum Theory* (New York: Prentice Hall)

Bohm D 1952a A suggested interpretation of quantum theory using 'hidden' variables. I *Phys. Rev.* **85** 166–79

Bohm D 1952b A suggested interpretation of quantum theory using 'hidden' variables. II *Phys. Rev.* **85** 180–93

Bohm D and Bub J 1966 A proposed solution of the measurement problem in quantum mechanics by hidden variable theory *Rev. Mod. Phys.* **38** 453–69

Born M 1949 *Natural Philosophy of Cause and Chance* (Oxford: Clarendon)

Casado A, Guerra S and Placido J 2008 Wigner representation for experiments on quantum cryptography using two-photon polarization entanglement produced in parametric down-conversion *J. Phys. B: At. Mol. Phys.* **41** 045501

Choi J, Oh S and Kim J 2020 Quantum approximation for wireless scheduling *Appl. Sci.* **10** 7116

Clauser J F, Horner M A, Shimony A and Holt R A 1969 Proposed experiment to test local hidden variable theories *Phys. Rev. Lett.* **23** 880–83

Davies P C W 2009 The quantum life *Phys. World* **22** 24–8

Davies P C W 2020 Does new physics lurk within living matter? *Phys. Today* **73** 34–40

Dirac P A M 1958 *The Principles of Quantum Mechanics* 4th edn (Oxford: Oxford University Press)

Duarte F J 1989 Optical device for rotating the polarization of a light beam *US Patent* 4822150

Duarte F J 1993 On a generalized interference equation and interferometric measurements *Opt. Commun.* **103** 8–14

Duarte F J 2002 Secure interferometric communications in free space *Opt. Commun.* **205** 313–9

Duarte F J 2003 *Tunable Laser Optics* 1st edn (New York: Elsevier Academic)

Duarte F J 2012 The origin of quantum entanglement experiments based on polarization measurements *Eur. Phys. J.* H **37** 311–8

Duarte F J 2014 *Quantum Optics for Engineers* (New York: CRC Press)

Duarte F J 2019 *Fundamentals of Quantum Entanglement* (Bristol: IOP Publishing)

Duarte F J and Paine D J 1989 Quantum mechanical description of *N*-slit interference phenomena *Proceedings of the International Conference on Lasers'88* ed R C Sze and F J Duarte (McLean, VA: STS Press) 552–54

Duarte F J, Taylor T S and Slaten J C 2020 On the probability amplitude of quantum entanglement and the Pauli matrices *Opt. Quantum Electron.* **52** 106

Einstein A, Podolsky B and Rosen N 1935 Can quantum mechanical description of physical reality be considered complete? *Phys. Rev.* **47** 777–80

Einstein A and Rosen N 1935 The particle problem in the general theory of relativity *Phys. Rev.* **48** 73–7

Ekert A K 1991 Quantum cryptography based on Bell's theorem *Phys. Rev. Lett.* **67** 661–63

Feynman R P 1985 Quantum mechanical computers *Opt. News* **11** 11–20

Feynman R P 1986 Quantum mechanical computers *Found. Phys.* **16** 507–31

Feynman R P, Leighton R B and Sands M 1965 *The Feynman Lectures on Physics* vol. III (Reading, MA: Addison-Wesley)

Fowles G R 1968 *Introduction to Modern Optics* (New York: Holt, Rinehart, and Winston)

Gisin N, Ribordy G, Tittel W and Zbinden H 2002 Quantum cryptography *Rev. Mod. Phys.* **74** 145–95

Jennewein T, Simon C, Weihs G, Weinfurter H and Zeilinger A 2000 Quantum cryptography with entangled photons *Phys. Rev. Lett.* **84** 4729–32

Kim Y-H, Kulik S P and Shih Y 2001 Quantum teleportation of a polarization state with a complete Bell state measurement *Phys. Rev. Lett.* **86** 1370–73

Kuroda T *et al* 2013 Symmetric quantum dots as efficient sources of highly entangled photons: violation of Bell's inequalities without spectral and temporal filtering *Phys. Rev.* B **88** 041306 (R)

Lee H, Kok P and Dowling J P 2002 A quantum Rosetta stone for interferometry *J. Mod. Opt.* **49** 2325–38

Ma X *et al* 2012 Quantum teleportation over 143 kilometres using active feed-forward *Nature* **489** 269–73

Maldacena J and Susskind L 2013 Cool horizons for entangled black holes *arXiv* 1306.0533v2

McFadden J and Al-Khalili J 2014 *Life on the Edge: The Coming Age of Quantum Biology* (New York: Broadway Books)

Misner C W and Wheeler J A 1957 Classical physics and geometry *Ann. Phys.* **2** 525–603

Misner C W, Thorne K S and Wheeler J A 2017 *Gravitation* (Princeton, NJ: Princeton University Press)

Naik D S, Peterson C G, White A G, Berglund A J and Kwiat P G 2000 Entangled state quantum cryptography: eavesdropping on the Ekert protocol *Phys. Rev. Lett.* **84** 4733–36

Nielsen M A and Chuang I L 2000 *Quantum Computation and Quantum Information* (Cambridge: Cambridge University Press)

Ou Z-Y J 2017 *Quantum Optics for Experimentalists* (Singapore: World Sceintific)

Pirandola S, Eisert J, Weedbrook C, Furusawa A and Braunstein S L 2015 Advances in quantum teleportation *Nat. Photon* **9** 641–52

Pryce M H L and Ward J C 1947 Angular correlation effects with annihilation radiation *Nature* **160** 435

Rauch D *et al* 2018 Cosmic Bell test using random measurement settings from high-redshift quasars *Phys. Rev. Lett.* **121** 080403

Rota M B, Basset F B, Tedeshi D and Trotta R 2020 Entanglement teleportation with photons from quantum dots: toward a solid-state based quantum network *IEEE J. Sel. Top. Quantum Electron.* **26** 6400416

Schnabel R, Mavalvala N, McClelland D E and Lam P K 2010 Quantum metrology for gravitational wave astronomy *Nat. Commun* **1** 121

Schrödinger E 1944 *What Is Life?* (Cambridge: Cambridge University Press)

Schumacher B 1995 Quantum coding *Phys. Rev. Lett.* **51** 2738–47

Steane A 1998 Quantum computing *Rep. Prog. Phys.* **61** 117–74

Tamburini F and Licata I 2020 Testing the equivalence principle and discreteness of spacetime through the t^3 gravitational phase with quantum information technology *Phys. Lett.* B **810** 135792

Taylor M A and Bowen W P 2016 Quantum metrology and its applications is biology *Phys. Rep.* **615** 1–59

Tittel W, Brendel J, Zbinden H and Gisin N 2000 Quantum cryptography using entangled photons in energy-time Bell states *Phys. Rev. Lett.* **84** 4737–40

Ursin R *et al* 2007 Free-space distribution of entanglement and single photons over 144 km *Nat. Phys.* **3** 481–86

Vadimov V, Ala-Nissila T and Möttönen M 2020 Persistent correlations in many-body localized spin chains *Phys. Rev. Res* **2** 043154

Villar A *et al* 2020 Entanglement demonstration on board a nano-satellite *Optica* **7** 734–37

Ward J C 1949 *Some Properties of the Elementary Particles* (Oxford: Oxford University Press)

Wiesner S 1983 Conjugate coding *ACM SIGACT News* **15** 78–88

Williams C P and Clearwater S H 1998 *Explorations in Quantum Computing* (Berlin: Springer)

Yin Y *et al* 2012 Quantum teleportation and entanglement distribution over 100-kilometre free-space channels *Nature* **488** 185–88

Chapter 9

Space-to-space quantum communications

Parameters relevant to satellite communications are reviewed, as well as existing data from satellite quantum entanglement experiments. Various satellite network concepts, of practical relevance, are considered.

9.1 Introduction

Among the first antecedents in the open literature on the use of quantum optics techniques in space-to-space secure communications are those of Duarte (2002) and Aspelmeyer *et al* (2003). Duarte (2002) focused on the use of N-slit quantum interferometric techniques while Aspelmeyer *et al* (2003) considered quantum entanglement techniques. Further discussions on the use of interferometric and quantum entanglement configurations in satellite-to-satellite communications were provided by Duarte and Taylor (2015) and Duarte (2016).

Sources for quantum entanglement free-space configurations can emit via the following probability amplitudes (Pryce and Ward 1947, Ward 1949, Duarte 2019)

$$|\psi\rangle_+ = 2^{-1/2}(|x\rangle_1|y\rangle_2 + |y\rangle_1|x\rangle_2) \tag{9.1}$$

$$|\psi\rangle_- = 2^{-1/2}(|x\rangle_1|y\rangle_2 - |y\rangle_1|x\rangle_2) \tag{9.2}$$

$$|\psi\rangle^+ = 2^{-1/2}(|x\rangle_1|x\rangle_2 + |y\rangle_1|y\rangle_2) \tag{9.3}$$

$$|\psi\rangle^- = 2^{-1/2}(|x\rangle_1|x\rangle_2 - |y\rangle_1|y\rangle_2) \tag{9.4}$$

Available quanta pair sources also emit in modified quantum entanglement states such as

$$|\psi\rangle_+ = 2^{-1/2}(|x\rangle_1|y\rangle_2 + e^{i\Delta\varphi}\,|y\rangle_1|x\rangle_2) \tag{9.5}$$

$$|\psi\rangle^+ = 2^{-1/2}(|x\rangle_1|x\rangle_2 + e^{i\Delta\varphi}\,|y\rangle_1|y\rangle_2) \tag{9.6}$$

doi:10.1088/978-0-7503-3407-5ch9

So far, the standard quantum key distribution (QKD) methodology being applied utilizes the Ekert (1991) protocol that relies on Bell's inequalities (Bell 1964) or modified Bell's inequalities (Clauser *et al* 1969) of the form

$$\Sigma_P = \left|P(\varphi_1, \varphi_2) - P(\varphi_1, \varphi_2') + P(\varphi_1', \varphi_2) + P(\varphi_1', \varphi_2')\right| \leqslant 2 \tag{9.7}$$

for security verification (see chapter 10).

9.2 Satellite engineering parameters

Space-to-space, satellite-to-satellite, and Earth–space–Earth optical communications are subjects deeply submerged in engineering techniques and engineering know-how.

Assuming that the physics works, the following engineering items, in alphabetical order, are significant:

- configurational compactness;
- detection efficiency;
- mechanical ruggedness;
- optical ruggedness;
- orbital altitude: either low Earth orbit (LEO: ~200–1000 km) or geostationary orbit (GEO: ~35 786 km);
- pointing and tracking errors;
- power consumption;
- propagation losses;
- radiation ruggedness;
- signal-to-noise ratio;
- system reliability;
- system weight.

The engineering and technology of some of these topics are discussed in detail in references such as Lambert and Casey (1995) and Kaushal and Kaddoum (2017). For instance, for Earth–space–Earth communications, propagation parameters of interest include the atmospheric transmittance $T(\lambda, \theta, h)$ and the attenuation due to fog $\beta(\lambda)$. At zenith angle θ

$$T(\lambda, \theta, h) = \exp - \left(\sec\theta \int_0^H \gamma(\lambda, h) dh\right) \tag{9.8}$$

where γ is the atmospheric attenuation, H is the vertical propagation height, and λ is the emission wavelength. For $\lambda = 550$ nm the fog attenuation from Mie scattering is given by

$$\beta(\lambda) = \left(\frac{3.91}{V}\right)\left(\frac{\lambda}{550}\right)^{-p} \tag{9.9}$$

where V is the visibility range (km), λ is the emission wavelength (nm), and p is the scattering size distribution coefficient (Kaushal and Kaddoum 2017). The beauty of space-to-space quantum communications is that the problems and difficulties associated with atmospheric propagation are obviated.

9.3 Beam divergence

Starting from Heisenberg's uncertainty principle

$$\Delta p \, \Delta x \approx h \tag{9.10}$$

and $p = \hbar k$, so that

$$\Delta p = \frac{h}{2\pi} \Delta k \tag{9.11}$$

then (Duarte 2003, 2014)

$$\Delta k \, \Delta x \approx 2\pi \tag{9.12}$$

For a beam propagating in the z-direction, for a small angle θ, $k_x \approx k\theta$ so that

$$\Delta k_x \approx k \Delta \theta \tag{9.13}$$

Substitution into equation (9.12) leads directly to (Duarte 2003, 2014)

$$\Delta \theta \approx \frac{\lambda}{\Delta x} \tag{9.14}$$

Beam divergence is directly proportional to wavelength and inversely proportional to the transverse cross-sectional dimension of the propagating beam. The detailed physics for *Gaussian* beam propagation is more subtle but the beauty is that for an optimally designed laser cavity emitting in a single-transverse-mode with a Gaussian profile equation (9.14) becomes (Duarte 1989, 2003, 2014)

$$\Delta \theta \approx \frac{\lambda}{\pi w_b} \tag{9.15}$$

where w_b is the beam waist at full-width-half-maximum (FWHM).

In general, near and far-field beam divergence calculations can be accurately carried out, without approximations, using the generalized interferometric probability for ensembles of indistinguishable photons, and its corresponding intensity equation (Duarte 1993, 2014, 2019)

$$\langle d|s\rangle\langle d|s\rangle^* = \sum_{j=1}^{N} \Psi(r_j)^2 + 2\sum_{j=1}^{N} \Psi(r_j)\left(\sum_{m=j+1}^{N} \Psi(r_m)\cos(\Omega_m - \Omega_j)\right) \tag{9.16}$$

$$\mathscr{I}(\nu) = r\,\mathrm{K}\,h\nu\langle d|s\rangle\langle d|s\rangle^* \tag{9.17}$$

This is done by representing a single aperture by a large number of N-imaginary equally spaced slits (Duarte 1993). In the near field the profile will be a structurized diffraction distribution and for the far field the propagation profile will assume a more familiar near-Gaussian or Gaussian profile (Duarte 1993, 2014). Indeed, the physics included in the interferometric equation (9.16) allows for the derivation of $\Delta p\, \Delta x \approx h$ and hence the divergence equation (9.14) (Duarte 2003, 2014).

Equation (9.16) either applies to a single photon ($r = 1$) or to a population of indistinguishable photons r. In the second case, r assumes the number of quanta available from the narrow-linewidth source, see chapter 2. For example, for a single-transverse-mode nearly diffraction-limited single-longitudinal-mode narrow-linewidth, $\Delta\nu \approx 350$ MHz, pulsed laser emitting near the limit imposed by the uncertainty principle ($\Delta\nu\, \Delta t \approx 1$), at $\lambda \approx 590$ nm, this number can be rather high $r \approx 3 \times 10^{14}$(Duarte 1999).

In order to illustrate these calculations let us consider a laser with the characteristics just given above that exhibits a nearly diffraction-limited beam divergence of $\Delta\theta \approx 2.2$ mrad for a beam waist of $2w \approx 250\mu$m. Over a distance of 500 km this beam spreads into an enormous wavefront thus severely decreasing the photon intensity density. In order to reduce the spread of the wavefront the experimenter can use beam expansion techniques at the source. For instance, using miniature telescopic beam pre-expansion followed by multiple-prism beam expansion (Duarte and Piper 1982, Duarte 2003) the beam can be magnified to, lets say, $M2w \approx 300$ mm and then reduced in width to 250 mm via the use of apertures. As soon as apertures are introduced the straightforward Gaussian propagation physics ceases to be applicable and the generalized interferometric equation (9.16) must be used (Duarte 1993). Representing the 250 mm slit by 250 slits each 500 μm wide separated by 500 μm the far field distribution becomes near Gaussian with a FWHM ~1.5 m. See chapter 10 for a graphical representation of this type of interferometric computation.

The effect of terrestrial air turbulence on the propagation of superposition probability intensity distributions has been studied by Duarte *et al* (2010, 2011).

9.4 Optical configuration for quantum satellite communications

As described by Duarte (2019) the usual satellite-to-satellite communication configuration, between satellite A and satellite B, is illustrated in figure 9.1, where A is portrayed as the emitter and B as the receiver. For two active satellite emitters the configuration is shown in figure 9.2.

A generic quantum entanglement configuration is depicted in figure 9.3 where S is the source and A and B become receivers. This configuration is an obvious variation of the first known diagram on quantum entanglement due to Pryce and Ward (1947) and Ward (1949). A linear satellite array is depicted in figure 9.4.

For space-to-Earth quantum communications the configuration is the same as in figure 9.3 with satellites A and B being replaced by receiving stations on the ground.

9.4.1 Satellite–satellite

The basic intrinsic optical configuration for a satellite-to-satellite quantum communications system is depicted in figure 9.5. At least two alternatives are possible here:

1. An asymmetrical arrangement where S and A are located on the same platform and B is distant.
2. A near-symmetrical arrangement where S is remote to A and B.

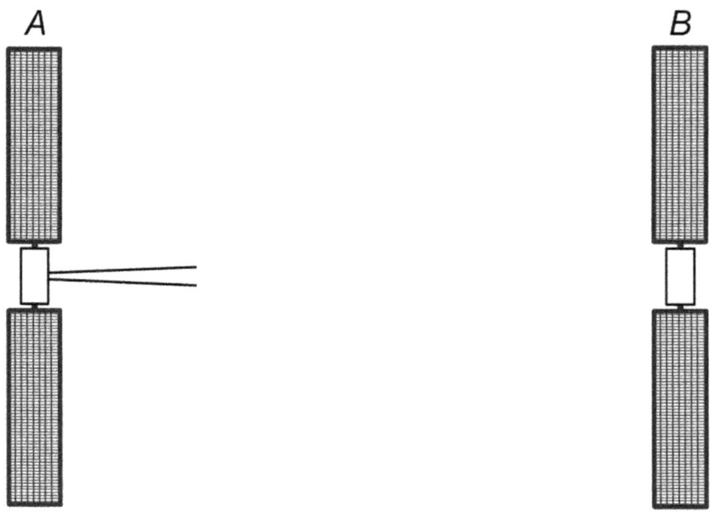

Figure 9.1. Conceptual schematics illustrating laser, or photon, communications between a transmitting satellite *A* and a receiving satellite *B* (Duarte 2019).

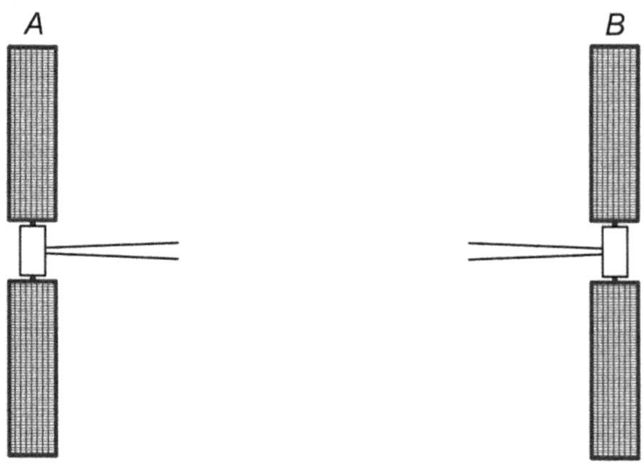

Figure 9.2. Conceptual schematics illustrating laser, or photon, communications between satellite *A* and a receiving satellite *B*. Here, both satellites are capable of transmission and reception (Duarte 2019).

Note that this diagram is configurationally accurate. In principle, with the proper quantum entanglement source, no additional optics should be necessary.

9.4.2 Earth–space–Earth

The basic intrinsic optical configuration for an Earth–satellite–Earth quantum communications system is depicted in figure 9.6. This is a near-symmetrical arrangement where *S* is remote to *A* and *B*.

Again, note that this diagram is configurationally accurate. In principle, with the proper quantum entanglement source, no additional optics should be necessary.

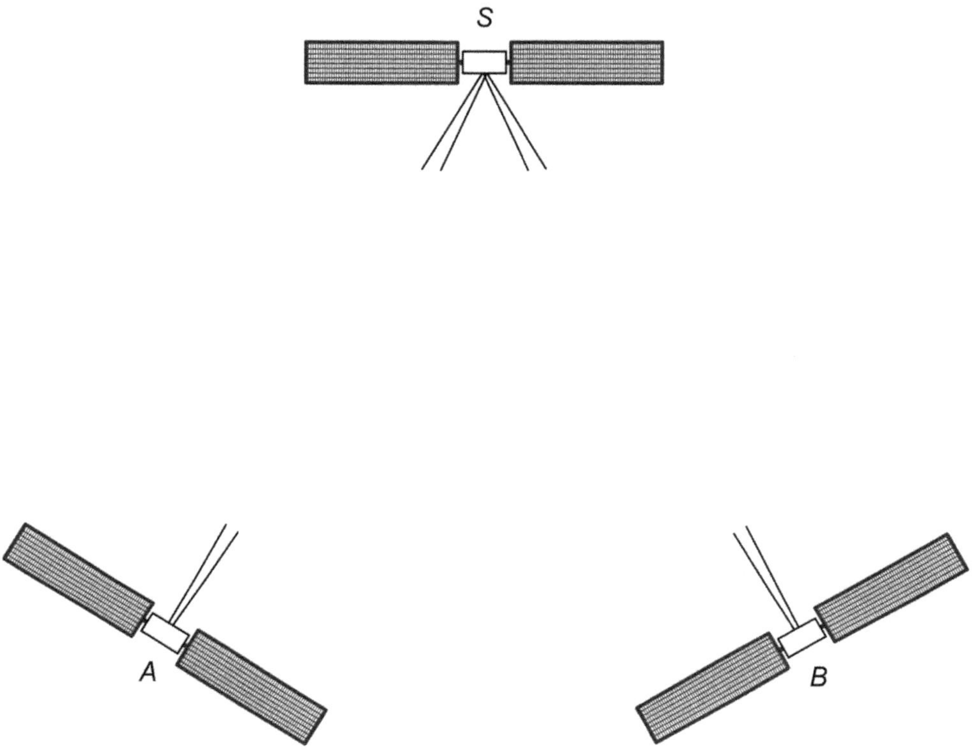

Figure 9.3. Conceptual schematics illustrating entangled photon communications between satellite *A* and satellite *B*. The source of the entangled photon pairs is at satellite *S* (Duarte 2019).

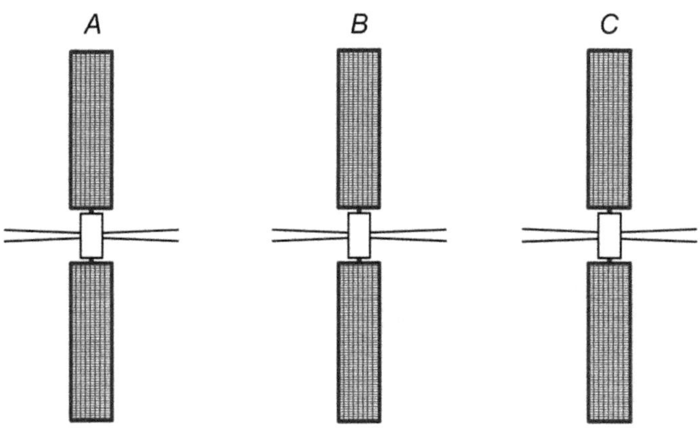

Figure 9.4. Conceptual schematics illustrating a satellite array at either LEO or GEO. Satellites have both transmitting and receiving capabilities (Duarte 2019).

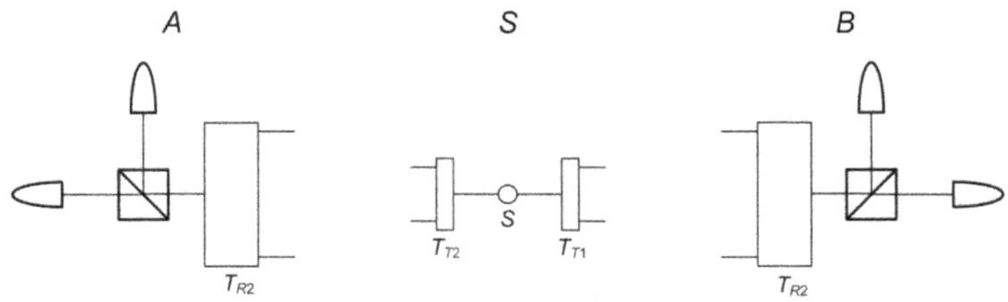

Figure 9.5. Conceptual schematics illustrating optimized optical configuration for space-to-space quantum entanglement communications. This optimized configuration includes the intrinsic source of polarization entangled photon pairs, transmission telescopes (T_{T1} and T_{T2}), receiving telescopes (T_{R1} and T_{R2}), in addition to the detection optics. These telescopes are reflection telescopes either Cassegrainian or modified Cassegrainians.

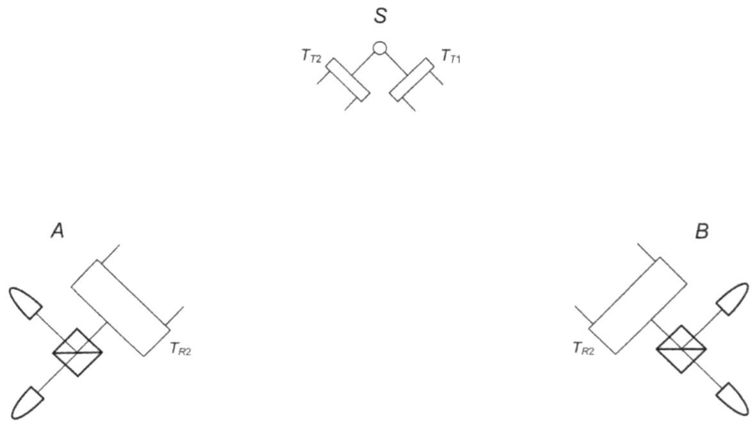

Figure 9.6. Conceptual schematics illustrating optimized optical configuration for Earth–Space–Earth quantum entanglement communications. This optimized configuration includes the intrinsic source of polarization entangled photon pairs, transmission telescopes (T_{T1} and T_{T2}), receiving telescopes (T_{R1} and T_{R2}), in addition to the detection optics. These telescopes are reflection telescopes either Cassegrainian or modified Cassegrainians. *A* and *B* are the terrestrial receiving stations.

9.5 Existing data from experiments on quantum satellite communications

In this section a summary of relevant parameters from Earth–space–Earth experiments published in the open literature is given. So far there is no data available in the open literature from space-to-space quantum communications experiments either utilizing quantum interferometry or quantum entanglement.

9.5.1 The Delingha–Lijiang experiment

The Delingha–Lijiang experiment (Yin *et al* 2012) took place with a photon source at an altitude of ~500 km. This experiment included some 130 optical and electro-optical parts. The emission source is a diode-laser pumped PPKTP crystal configured within a Sagnac interferometric ring. The CW pump diode laser emits a power of 30 mW at $\lambda \approx 405$ nm with a linewidth of $\Delta\nu \approx 160$ MHz. The type II spontaneous parametric down conversion (SPDC) emission is reported as 5.9×10^6 photon pairs per second at $\lambda \approx 810$ nm. Yin *et al* (2012) indicate that this type II SPDC source produces entangled quanta pairs described by the probability amplitude

$$|\psi\rangle_+ = 2^{-1/2}(|x\rangle_1|y\rangle_2 + |y\rangle_1|x\rangle_2) \tag{9.18}$$

It is noteworthy to observe that following the 810 nm quanta pair source there is an optical set composed of QWP–QWP–HWP plates used 'for polarization compensation' (Yin *et al* 2012). An additional set of QWP–QWP–HWP plates is installed prior to the entrance of both receiving telescopes on Earth. Yin *et al* (2012) report quantum key distribution while violating the modified Bell's inequalities (Clauser *et al* 1969) with a $\Sigma_{|\psi\rangle|\psi\rangle^*} = 2.37 \pm 0.09$.

9.5.2 The International Space Station experiment

A 'nanosatellite' in-orbit quantum entanglement experiment launched from the International Space Station (Villar *et al* 2020) took place at an altitude of ~408 km.

This experiment included only about 13 optical and electro-optical parts. The emission source is a diode-laser pumped BBO configuration in series that yields pairs of quanta with orthogonal polarizations which are nearly collinear. The CW pump diode laser emits a power of 17 mW at $\lambda \approx 405$ nm with a linewidth of $\Delta\nu \approx 160$ MHz. The type I spontaneous parametric down conversion (SPDC) emission is reported as 5.9×10^6 photon pairs per second at $\lambda \approx 800$ nm. Villar *et al* (2020) indicate that this type I SPDC source produces entangled quanta pairs described by the probability amplitude

$$|\psi\rangle^+ = 2^{-1/2}(|x\rangle_1|x\rangle_2 + e^{i\Delta\varphi}\,|y\rangle_1|y\rangle_2) \tag{9.19}$$

where $\Delta\varphi$ is the relative phase difference of quanta emitted from BB-1 and BBO-2.

Villar *et al* (2020) report quantum key distribution with a violation of the modified Bell's inequalities (Clauser *et al* 1969) at $\Sigma_{|\psi\rangle|\psi\rangle^*} = 2.60 \pm 0.06$.

The overall volume of this experiment was approximately $300 \times 85 \times 49$ mm at a weight of ~ 2.6 kg. Peak power consumption is reported at 2.5 W (Villar *et al* 2020).

9.6 Satellite networks and their dependence on entangled photon source characteristics

Consider the satellite-to-satellite configuration illustrated in figures 9.4 and 9.5 where each satellite has the components necessary for quantum entangled photon transmission and reception. Figures 9.7 and 9.8 give a more detailed configuration for this type of space-to-space entanglement communications system. Figure 9.7 shows the configuration for transmitting a photon from an entangled photon pair source to a receiving satellite some distance, D_{ss}, away. The transmitting satellite has a QE source typically as described previously in this chapter. Photons from the entangled pair source need to be sent on separate optical paths. One of the photons remains onboard the transmitting satellite and is kept in an optical delay path that needs to be configured by active optical elements that must 'hold' the photon in a

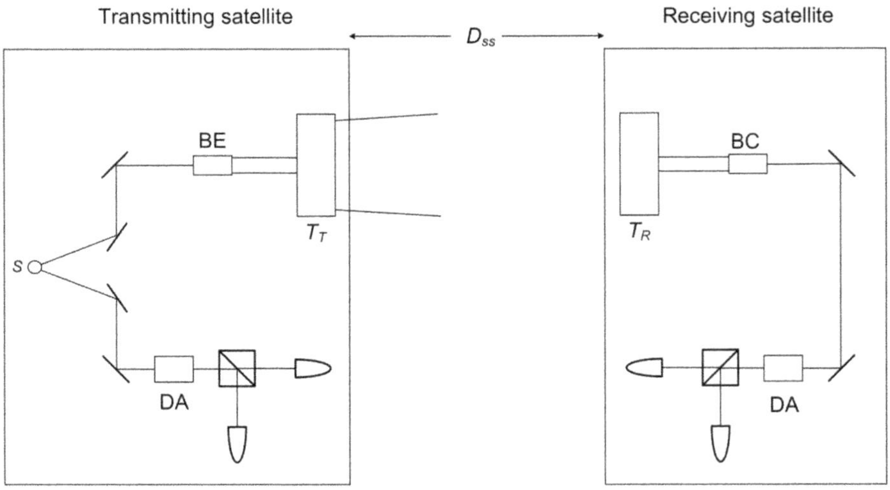

Figure 9.7. Conceptual minimal satellite configuration for transmitting a photon from an entangled photon pair source to a receiving satellite some distance, D_{SS}, away. BC: beam compressor, BE: beam expander, DA: delay adjuster, T_T: transmission telescope, T_R: receiving telescope.

delay loop (DL) until its entangled counterpart makes it way to the receiver satellite components. This delay loop function could also be implemented through some sort of 'quantum memory' device (Cho *et al* 2016, Ding *et al* 2015) or with an optical synchronization pulse scheme (Makino *et al* 2016).

The counterpart entangled photon travels through beam expanding and collimating transmission optics, typically a telescope assembly, then across space to the receiver. The receiver needs an optical telescope and beam reducing optics to transfer the received entangled photon to an optical path delay adjusting active optic, or quantum memory device, and then to the Bell state measuring apparatus. Figure 9.8 shows both transmitting and receiving systems on each satellite which are necessary for each satellite in a multiple spacecraft constellation. Note that this type of satellite-to-satellite configuration is for EPR type entanglement measurements implementing the standard Pryce–Ward entanglement equation as discussed throughout this book.

A practical and very key limitation of satellite-to-satellite quantum entanglement-based communications is the low light detection requirements. As discussed in section 9.5.1, a type of QE source is a type II SPDC crystal to generate the entangled pairs at a rate of about 5.9×10^6 photon pairs per second. The beam is split and therefore 5.9×10^6 single photons per second are then transmitted through the optical system, across space to the receiving optics, and then into a photon counting detection apparatus.

Each of these photons are entangled with a counterpart held back on the transmitting satellite. In the case of the Delingha–Lijiang experiment the counterpart entangled photons were actually transmitted to the Earth's surface.

Considering an idealized (nonrealizable) system with no optical losses other than those from diffraction alone demonstrates the difficulty of creating space-to-space systems. In reality, there would be scattering and absorption within the optical system.

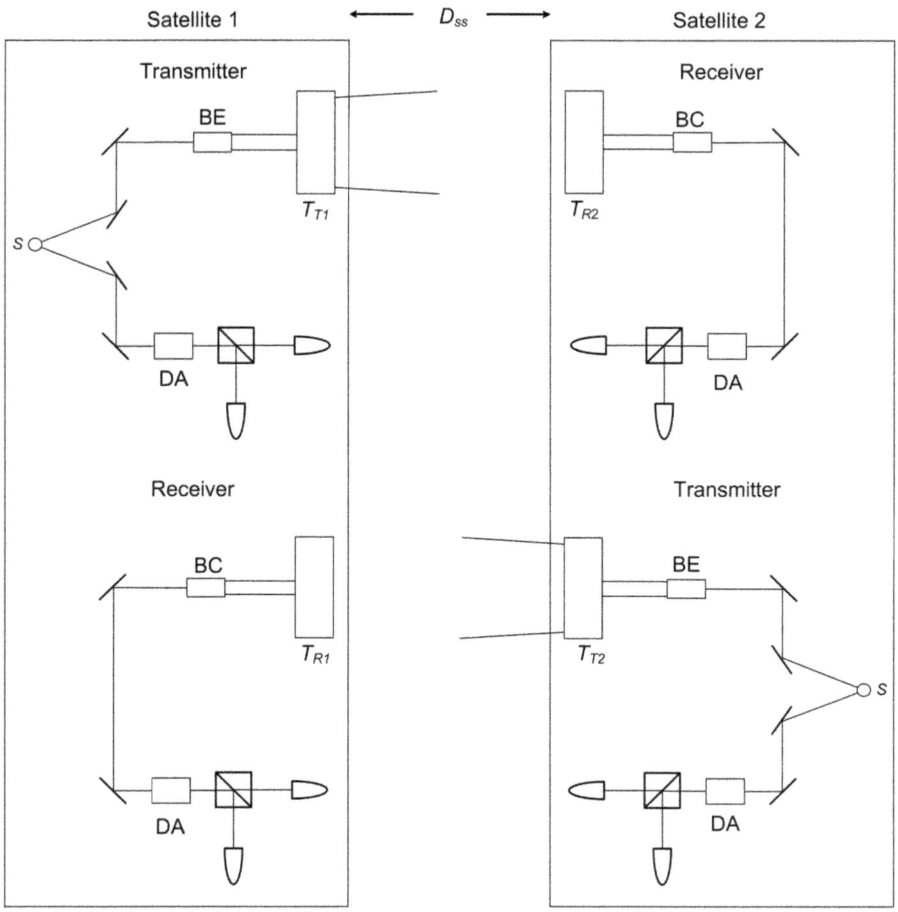

Figure 9.8. Conceptual minimal satellite configuration for space-to-space entanglement communications illustrating transmitter satellite and receiver satellite separated by D_{SS}. BC: beam compressor, BE: beam expander, DA: delay adjuster, T_{T1} and T_{T2}: transmission telescopes, T_{R1} and T_{R2}: receiving telescopes.

There would be losses due to detection error. There would be photon capture losses due to pointing jitter into the system and there would likely be background signal noise due to stray light sources. For this discussion all of these, and many other possible sources of error, are neglected.

Consider a hypothetical satellite constellation in LEO at a 500 km altitude in a circular orbit as shown in figure 9.9. The constellation consists of an arbitrary number, N_s, of satellites. Therefore, the distance between the satellites is determined by the chord length between them

$$D_{ss} = 2z \sin\left(\frac{\pi}{N_s}\right) \tag{9.20}$$

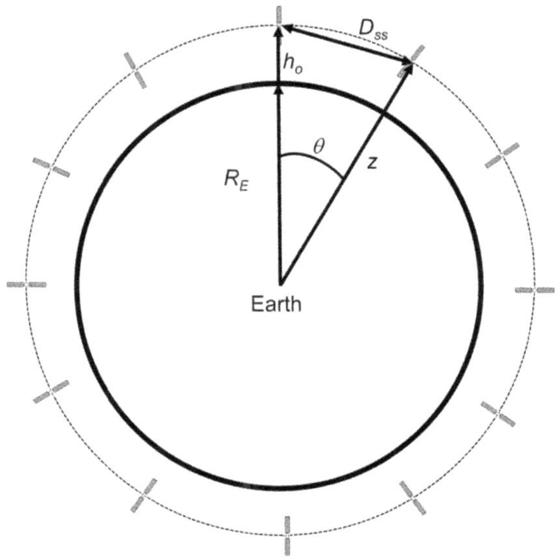

Figure 9.9. Conceptual satellite constellation in LEO at a 500 km altitude in a circular orbit. The figure is not drawn to scale.

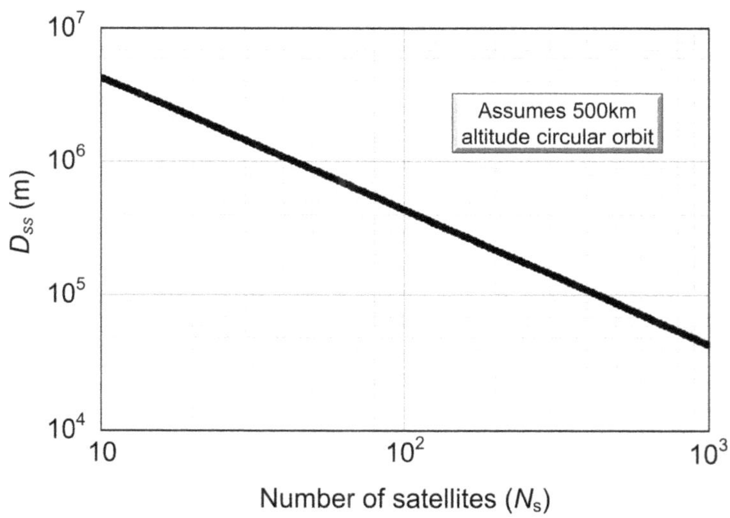

Figure 9.10. The distance between the satellites, D_{SS}, as a function of the number of satellites in the orbit determined by equation (9.20).

where

$$z = R_E + h_o \tag{9.21}$$

Here, R_E is the radius of the Earth and h_o is the orbit altitude from the surface of the Earth. Figure 9.10 shows the distance between the satellites, D_{SS}, as a function of the number of satellites in the orbit as determined by equation (9.20).

Assuming that the optical telescope for the satellite transmitter and receiver implement standard circularly symmetric optics, such as a Cassegrainian, Ritchey–Chretien, Newtonian, Galilean, Keplerian, or similar telescope, with circular apertures then the diameter d_{eb} of the entangled photon beam at the receiver aperture distance is calculated as

$$d_{eb} = \frac{2.44\lambda D_{ss}}{r_1} \tag{9.22}$$

Here, λ is the wavelength of the entangled photon beam and r_1 is the radius of the transmitting aperture. Equation (9.22) shows that the larger the distance between the satellites the larger the spot size at the receiving aperture and therefore the entangled photon beam number of photons will be spread over a larger area which is entirely consistent with the propagation physics discussed in section 9.3. This implies that a large aperture is required to gather as many of the entangled photons as possible to improve the probability of detection and, likewise, received data rate. The number of photons per second (n_r/t) in the beam captured by the receiver aperture of radius, r_2, is

$$\frac{n_r}{t} = \frac{n_t}{t}\frac{r_2^2}{d_{eb}^2} \tag{9.23}$$

which can also be expressed as

$$\frac{n_r}{t} = \frac{n_t}{t} r_1^2 r_2^2 \left(2.44\lambda z \sin\left(\frac{\pi}{N_s}\right)\right)^{-2} \tag{9.24}$$

Again, neglecting any losses internal to the optical system, equation (9.24) provides an acceptable design tool for approximately determining the transmitter and receiver aperture sizes based on the number of satellites in the constellation. Assuming an entangled pair source delivering 5.9×10^6 photon pairs per second at $\lambda \approx 800$ nm (Yin *et al* 2012, Villar *et al* 2020), and satellite-to-satellite connectivity, figures 9.11 through 9.13 show aperture size requirements versus the number of photons detected (n_d) as well as the number of satellites per constellation. Figure 9.11 simulates a ten-satellite constellation and shows that apertures must be considerably large in order to detect *all* of the entangled photons. In fact, the minimum radii for maximum detection are $r_1 = 0.2$ m and $r_2 = 10$ m. A more realistic design choice would be $r_1 = 1.0$ m and $r_2 = 2.0$ m, however, these are still fairly large optical systems for spacecraft. For comparison, the Hubble Space Telescope aperture radius is 1.2 m (NASA 2021). Note that smaller apertures could be used but figure 9.11 suggests that the number of photons per second detected would be much lower. Using $r_1 = r_2 = 0.1$ m apertures on both satellites would only allow for about 100 photons per second to be detected with an ideal system. The data rate with such a system would be extremely limited. Figure 9.12 shows a simulation of a constellation of 100 satellites. With this many satellites in the constellation there are several design choices that appear to be realizable. The $r_1 = r_2 = 0.1$ m apertures on each end of the link would enable a detection of over 10^4 photons per second which might be enough

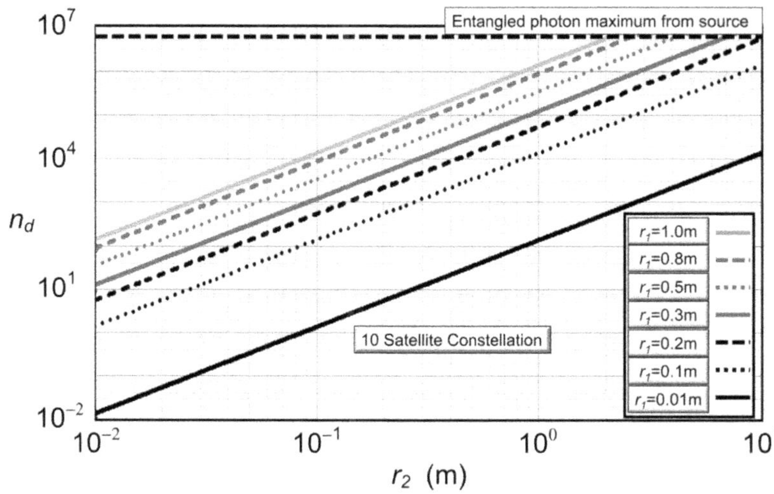

Figure 9.11. Number of detected photons n_d as a function of r_2 for a $N_s = 10$ satellite constellation assuming a quanta pair source yielding ~6×10^6 photons at $\lambda \approx 800$ nm. This simulation, at $N_s = 10$, suggests that the apertures radii must be considerably large in order to detect *all* of the emitted entangled photons between satellites configured as outlined in figure 9.9.

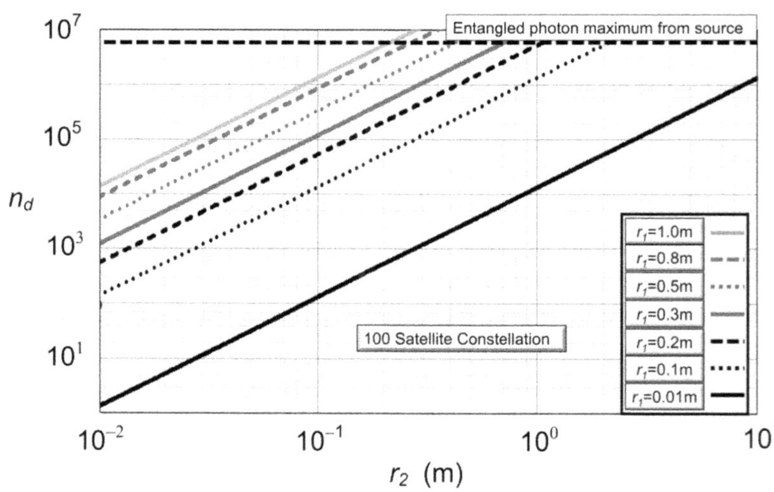

Figure 9.12. Number of detected photons n_d as a function of r_2 for a $N_s = 100$ satellite constellation assuming a quanta pair source yielding ~6×10^6 photons at $\lambda \approx 800$ nm.

throughput for useful data transmission. Figure 9.13 shows a simulation of a constellation with 1000 satellites. The $r_1 = r_2 = 0.1$ m apertures in this case enable over a million photons per second to be detected.

The situation can be improved rather significantly using brighter sources at shorter wavelengths. Assuming a future source emitting ~10^9 photons in the visible at $\lambda \approx 590$ nm the equivalent case performance for $N_s = 10$, $N_s = 100$, and $N_s = 1000$ is

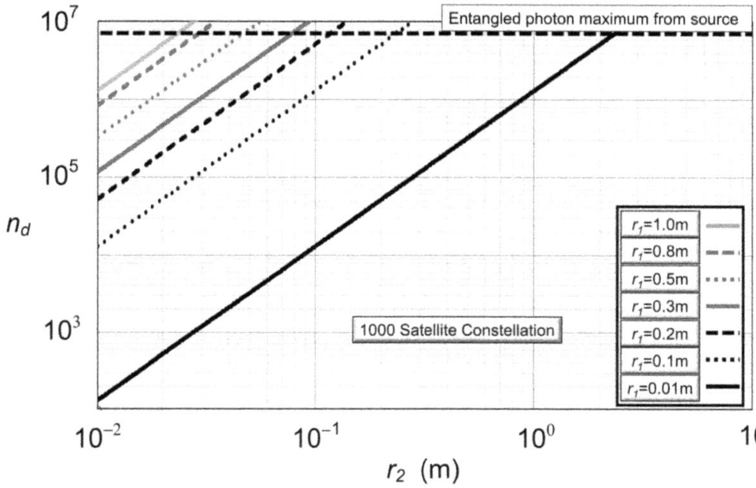

Figure 9.13. Number of detected photons n_d as a function of r_2 for a $N_s = 1000$ satellite constellation assuming a quanta pair source yielding ~6×10^6 photons at $\lambda \approx 800$ nm.

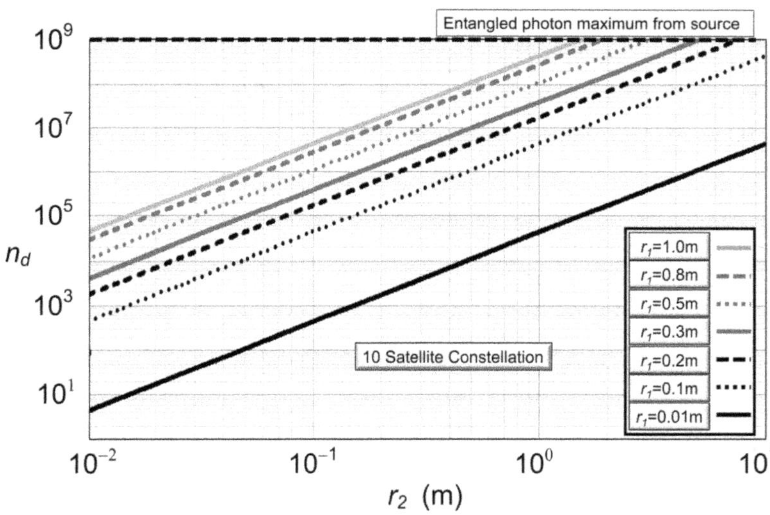

Figure 9.14. Number of detected photons n_d as a function of r_2 for a $N_s = 10$ satellite constellation assuming a quanta pair source yielding ~10^9 photons at $\lambda \approx 590$ nm.

illustrated in figures 9.14–9.16. For $N_s = 100$, with the brighter source a configuration with $r_1 = r_2 = 0.1$ m would allow the detection of ~10^7 photons per second as compared with ~10^4 using the dimmer longer wavelength option.

This conjectural discussion illuminates the need for the development of bright quantum entanglement sources. Any satellite-to-satellite quantum communications channel will need to transfer between them as many of the entangled photons generated as possible to minimize errors. As the satellite networks are implemented, the orbit altitude will also exacerbate the issue. The altitude of the orbit, if larger,

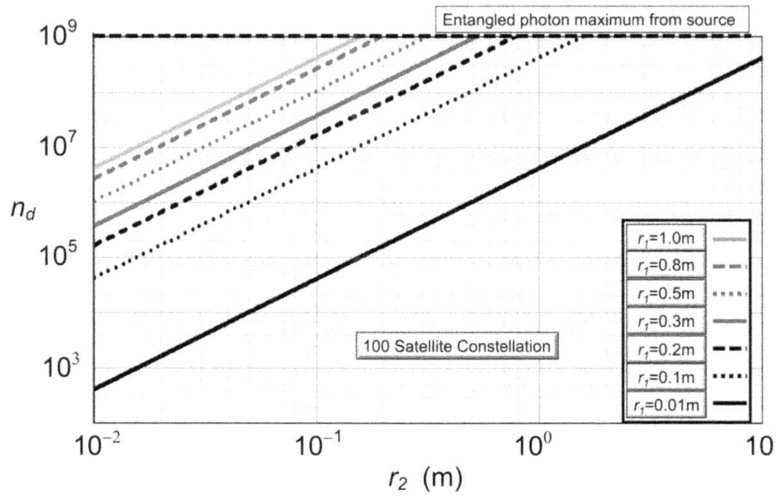

Figure 9.15. Number of detected photons n_d as a function of r_2 for a $N_s = 100$ satellite constellation assuming a quanta pair source yielding ~10^9 photons at $\lambda \approx 590$ nm.

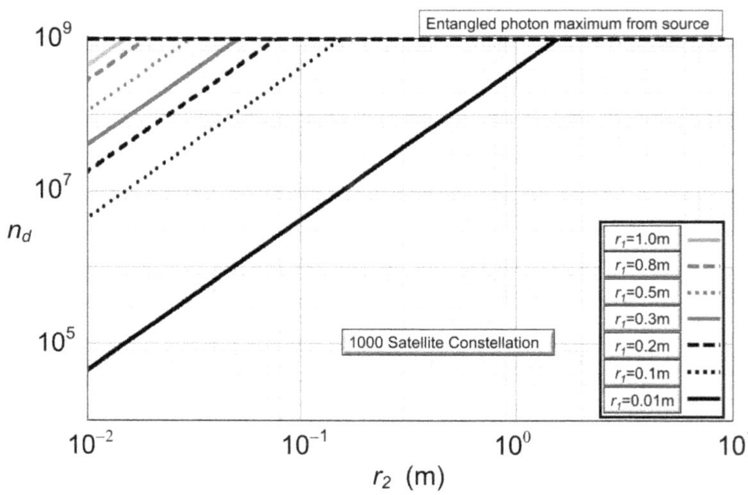

Figure 9.16. Number of detected photons n_d as a function of r_2 for a $N_s = 1000$ satellite constellation assuming a quanta pair source yielding ~10^9 photons at $\lambda \approx 590$ nm.

will mean that the distance between the satellites will be larger, which will increase the need for either larger apertures, more satellites, or brighter entanglement sources. More satellites and larger apertures increase logistics, technical complexity, and maintenance costs and will push the concept beyond cost effectiveness. The advantages of using brighter sources at shorter wavelengths, thus minimizing beam divergence, is self-evident. In the next section a summary of currently available photon pair sources is provided. Brighter photon pair sources, as considered above, should become available within a decade or so.

9.6.1 The quantum repeater and quantum data teleportation

The satellite-to-satellite quantum communications scenario discussed above only considers a basic EPR type transmission of data through entanglement between two points. A satellite network or constellation around the planet will undoubtedly require many satellites as illustrated in figure 9.9 in order to make the actual satellite design practical. However, the previous configuration discussion only considers transfer of data between two satellites. In modern day classical communications information is often sent from one system to the next and then forwarded on to the next system and so on.

Forwarding of a received electronic message is a real-world common example of this type of communications. Standard communication systems use 'repeaters' to receive and then forward the information to the next device. Classically, the information is simply recorded and retransmitted or 'routed' to its final destination.

A quantum communications network would need special devices to perform this 'routing' action of entangled information. The physics required is that of entanglement-based quantum teleportation, as discussed in section 8.4.1 and summarized in figure 8.7.

Consider figure 9.17 where sender A is located at satellite 1 generating an entangled photon pair 1&2, and sender B is located at satellite 2 generating entangled photons 3&4. Assume that sender A wants to teleport the following probability amplitude or state

$$|q\rangle_d = 2^{-1/2}(\alpha|1\rangle + \beta|0\rangle) \qquad (9.25)$$

The hypothetical satellite network depicted in figure 9.17 was outlined in order to enable this data to be transmitted, in fact teleported, from A at satellite 1 to B at satellite 2 exploiting the features of the entangled photons generated at each location. Photon 1 stays with A and photon 4 stays with B, however A sends 2 and B sends 3 to a central satellite that will do an entanglement measurement procedure via a Bell state measurement (BSM) device and other quantum processing.

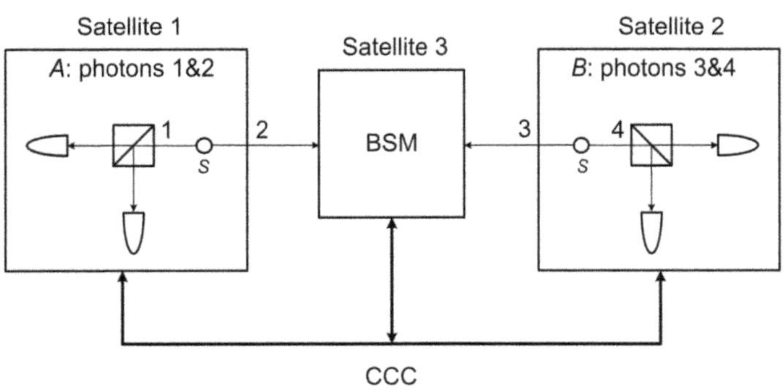

Figure 9.17. Sender A is located at satellite 1 and generates the entangled photon pair 1&2. Sender B is located at satellite 2 and generates entangled photons 3&4. A has a qbit of data to send to B and doesn't know the state of this qbit. BSM: Bell state measurement, CCC: classical communications channel.

Assume that the state of the photons 2 and 3 at the BSM satellite are processed in such a way that they become

$$|\psi\rangle_{14} = 2^{-1/2}(|1\rangle_1|1\rangle_4 + |0\rangle_1|0\rangle_4) \tag{9.26}$$

Equations (9.25) and (9.26) represent three qbits. A holds the qbit in equation (9.25) (i.e., $|q\rangle_d$, the data) and the qbit represented by equation (9.26) (the entangled photons 1 and 4 now connected because 2 and 3 went through a BSM process in the middle satellite). B holds the same qbit given by equation (9.26). This three qbit system is described by

$$|q\rangle_d|\psi\rangle_{14} = 2^{-1}(\alpha|1\rangle + \beta|0\rangle)(|1\rangle|1\rangle + |0\rangle|0\rangle) \tag{9.27}$$

$$|q\rangle_d|\psi\rangle_{14} = 2^{-1}(\alpha|111\rangle + \alpha|100\rangle + \beta|011\rangle + \beta|000\rangle) \tag{9.28}$$

The system is now set for quantum communication of the original data qbit. The mechanics of the process is described in chapter 8. The quantum state system to be implemented is

$$2^{-1}(|r\rangle_{14+}(\alpha|1\rangle + \beta|0\rangle) + |r\rangle_{14-}(\alpha|1\rangle - \beta|0\rangle) + |s\rangle_{14+}(\alpha|0\rangle + \beta|1\rangle) + |s\rangle_{14-}(\alpha|1\rangle - \beta|0\rangle)) \tag{9.29}$$

where

$$|s\rangle_{14+} = 2^{-1/2}(|1\rangle_1|0\rangle_4 + |0\rangle_1|1\rangle_4) \tag{9.30}$$

$$|s\rangle_{14-} = 2^{-1/2}(|1\rangle_1|0\rangle_4 - |0\rangle_1|1\rangle_4) \tag{9.31}$$

$$|r\rangle_{14+} = 2^{-1/2}(|1\rangle_1|1\rangle_4 + |0\rangle_1|0\rangle_4) \tag{9.32}$$

$$|r\rangle_{14-} = 2^{-1/2}(|1\rangle_1|1\rangle_4 - |0\rangle_1|0\rangle_4) \tag{9.33}$$

The sending device at A's location then makes a measurement to determine which of the four qbits is present. Based on the probability measured at A the basis state for B is determined. In other words, if A measures $|r\rangle_{14+}$ the base state is $(\alpha|1\rangle + \beta|0\rangle)$ similarly, for $|r\rangle_{14-}$ the base state is $(\alpha|1\rangle - \beta|0\rangle)$, for $|s\rangle_{14+}$ the base state is $(\alpha|0\rangle + \beta|1\rangle)$, and for $|s\rangle_{14-}$ the base state is $(\alpha|1\rangle - \beta|0\rangle)$.

The A device can now send to B's device, via a classical channel, the unitary transformation to be applied in order to recover $|q\rangle_d$. The qbit at the B location will be one of the four basis states listed above. B's receiving device can now implement the unitary transformation sent by A to recover $|q\rangle_d$:

$$|r\rangle_{14+} \rightarrow (\alpha|1\rangle + \beta|0\rangle) \rightarrow I(\alpha|1\rangle + \beta|0\rangle) \rightarrow (\alpha|1\rangle + \beta|0\rangle) \tag{9.34}$$

$$|r\rangle_{14-} \rightarrow (\alpha|1\rangle - \beta|0\rangle) \rightarrow \sigma_z(\alpha|1\rangle - \beta|0\rangle) \rightarrow (\alpha|1\rangle + \beta|0\rangle) \tag{9.35}$$

$$|s\rangle_{14+} \rightarrow (\alpha|0\rangle + \beta|1\rangle) \rightarrow \sigma_x(\alpha|0\rangle + \beta|1\rangle) \rightarrow (\alpha|1\rangle + \beta|0\rangle) \tag{9.36}$$

$$|s\rangle_{14-} \rightarrow (\alpha|0\rangle - \beta|1\rangle) \rightarrow i\sigma_y(\alpha|0\rangle - \beta|1\rangle) \rightarrow (\alpha|1\rangle + \beta|0\rangle) \tag{9.37}$$

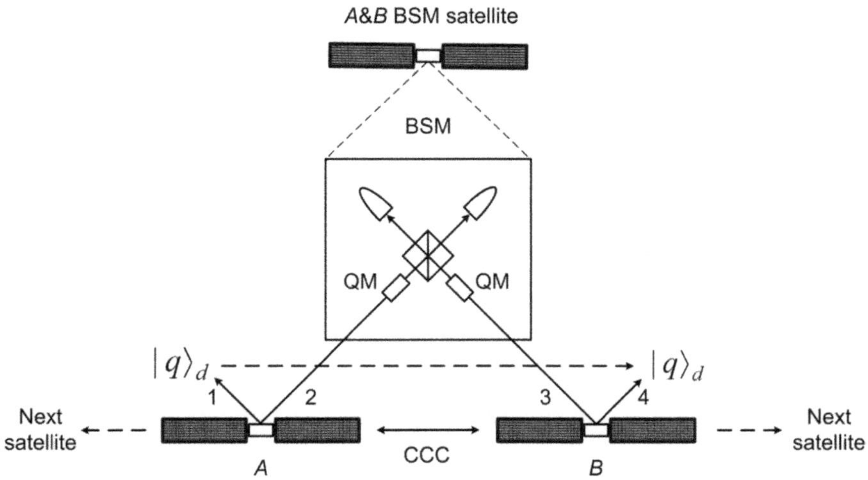

Figure 9.18. Conceptual outline for data teleportation in a satellite network. BSM: Bell state measurement, CCC: classical communications channel, QM: quantum memory.

where I, σ_x, σ_z, and $i\sigma_y$ are the unitary transformations (see chapter 8). Following one of the four operations in equations (9.34)–(9.37) the receiving device at B's location will then have the data qbit from A's device, that is, the original $|q\rangle_d$. Figure 9.18 outlines a pictorial implementation of this concept of data teleportation in a future satellite network, whereas a constellation of satellites would be required. This network should be integrated by A-type sending stations, B-type receiving stations, and BSM type relay stations. The components for these three required stations could likely be collocated on each satellite with appropriate transmitting and receiving optics. Local on-board control computer processing would determine which protocol or mode the satellite would operate at any given time.

9.7 Sources for quantum entanglement communications

Sources applicable to quantum communications, quantum cryptography, and quantum teleportation are listed next:

1. Type I spontaneous parametric down conversion (SPDC): these use optical pumping, at $\lambda \approx 266$ nm, of KD* P crystals which emit at $\lambda \approx 532$ nm (Shih and Alley 1988) and diode-laser excitation, at $\lambda \approx 405$ nm, of BBO crystals emitting at $\lambda \approx 800$ nm (Villar *et al* 2020).
2. Type II SPDC: these sources were first demonstrated using an optically pumped BBO crystal emitting at $\lambda \approx 702$ nm (Kwiat *et al* 1995). An alternative version of this scheme utilizes diode laser pumping, at $\lambda \approx 405$ nm, of a PPKTP crystal emitting at $\lambda \approx 810$ nm (Yin *et al* 2012).
3. Quantum dots: *biexiton* devices yielding entangled photon pairs at $\lambda \approx 790$ nm (Huber *et al* 2017), $\lambda \approx 873$ nm (Müller *et al* 2014), and $\lambda \approx 885$ nm (Salter *et al* 2010) have been reported. There is also research to generate

indistinguishable photons from multiple quantum dots (Singh *et al* 2019). The biexiton quantum entanglement emission is reported to be of the $|\psi\rangle^{+}$ form (Müller *et al* 2014).

The use of direct visible laser sources to randomly generate entangled photons with orthogonal polarization, propagating in opposite directions along the same optical axis, has been discussed by Duarte (2018, 2019). A Raman based device to generate randomly oriented linear polarizations, at $\lambda \approx 573$ nm, has been reported by Little *et al* (2021).

9.8 Outlook

The concept as well as the implementation of a quantum communications network is developing one experiment at a time. Any future network that will let quantum devices exchange information via the laws of quantum mechanics will require new protocols, technologies, and techniques for its implementation. In theory, a quantum internet would lend unprecedented capabilities that are impossible to carry out with today's web applications. Currently, quantum processing capabilities exist in a very small number of processors and incipient protocols. Standard qbits are typically modeled as two-bit systems based on the standard entanglement sources being based on two-level quantum phenomena. However, there are other quantum phenomena of multiple, *d*, levels. These types of phenomena could enable 'qdits' rather than 'qbits' where the '*d*' represents the number of levels of the quantum process and therefore increases the resolution and capability of the quantum channel (see chapter 2).

A constellation of many satellites, aerial systems, ground systems, and maritime systems is the basic concept design for the future of quantum communications. A quantum network with a few quantum processors at each device might manifest over time in a very large array of connected quantum processors analogous to the vast number of connected classical processors linked via the internet. Such a quantum network will have much broader implications because these systems, once interconnected, will be a global net of quantum processing. The entire network might be implemented as a singular quantum computer of many quantum processors creating an unprecedented powerful computing network that will enable the solution of problems currently beyond the abilities of any single quantum computer architecture.

Either for space-to-space or Earth–space–Earth communications, from an engineering perspective the main prerogative is to maintain the optical and electro-optical systems as close as possible to their intrinsic minimal configurations with a reduced number of optical and electro-optical elements. Better understanding of the physics is the preamble of engineering improvements. Equally important is the development of shorter wavelength high-brightness quantum entanglement sources.

Problems

9.1 Verify that for a beam propagating in the z-direction the approximation $k_x \approx k\theta$ is valid for small angles.

9.2 Assuming a Gaussian beam profile, and $2w \approx 250\ \mu\text{m}$, estimate using equation (9.15) the wavefront width for a propagation distance of 500 km.
9.3 Assuming a Gaussian beam profile, $2w \approx 1000\ \mu\text{m}$ and $\lambda \approx 590$ nm, estimate using equation (9.15), the propagation distance from the source at which the initial intensity per unit area (i.e. m^2) decreases by 99%.
9.4 Assuming equal initial intensities in J $s^{-1}m^{-2}$ and equal beam waist characteristics from two emission sources, one at $\lambda \approx 590$ nm and the other at $\lambda \approx 800$ nm, use $\Delta\theta \approx \lambda/\pi w$ to estimate the difference in intensities per unit area (i.e. m^2) at a distance of 100 km. Assume for both sources a Gaussian beam profile.
9.5 Outline the design of an Earth–Space–Earth quantum entanglement communications system with a minimum of optical and electro-optical parts.
9.6 Outline the design of a space-to-space quantum entanglement communications system with a minimum of optical and electro-optical parts.

References

Aspelmeyer M, Jennewein T, Pfennigbauer M, Leeb W R and Zeilinger A 2003 Long-distance quantum communication with entangled photons using satellites *IEEE J Sel. Top. Quantum Electron.* **9** 1541–51

Bell J S 1964 On the Einstein–Podolsky–Rosen paradox *Physics* **1** 195–200

Cho Y-W, Campbell G T, Everett J L, Bernu J, Higginbottom D B, Cao M T, Geng J, Robins N P, Lam P K and Buchler B C 2016 Highly efficient optical quantum memory with long coherence time in cold atoms *Optica* **3** 100–7

Clauser J F, Horner M A, Shimony A and Holt R A 1969 Proposed experiment to test local hidden variable theories *Phys. Rev. Lett.* **23** 880–83

Ding D S, Zhang W, Zhou Z Y, Shi S, Xiang G Y, Wang X S, Jiang Y K, Shi B S and Guo G C 2015 Quantum storage of orbital angular momentum entanglement in an atomic ensemble *Phys. Rev. Lett.* **114** 050502

Duarte F J 1989 Ray transfer matrix analysis of multiple-prism dye laser oscillators *Opt. Quantum Electron.* **21** 47–54

Duarte F J 1993 On a generalized interference equation and interferometric measurements *Opt. Commun.* **103** 8–14

Duarte F J 1999 Multiple-prism grating solid-state dye laser oscillator: optimized architecture *Appl. Opt.* **38** 6347–9

Duarte F J 2002 Secure interferometric communications in free space *Opt. Commun.* **205** 313–19

Duarte F J 2003 *Tunable Laser Optics* (New York: Elsevier Academic)

Duarte F J 2014 *Quantum Optics for Engineers* (New York: CRC Press)

Duarte F J 2016 Secure space-to-space interferometric communications and its nexus to the physics of quantum entanglement *Appl. Phys. Rev.* **3** 041301

Duarte F J 2018 Organic lasers for *N*-channel quantum entanglement *Organic Lasers and Organic Photonics* ed F J Duarte (Bristol: IOP Publishing) ch 15

Duarte F J 2019 *Fundamentals of Quantum Entanglement* (Bristol: IOP Publishing)

Duarte F J and Piper J A 1982 Dispersion theory of multiple-prism beam expander for pulsed dye lasers *Opt. Commun.* **43** 303–7

Duarte F J and Taylor T S 2015 Quantum entanglement physics secures space-to-space interferometric communications *Laser Focus World* **51** 54–8

Duarte F J, Taylor T S, Black A B, Davenport W E and Varmette P G 2011 *N*-slit interferometer for secure optical communications: 527 m intra interferometric path length *J. Opt.* **13** 035710

Duarte F J, Taylor T S, Clark A B and Davenport W E 2010 The *N*-slit interferometer: an extended configuration *J. Opt.* **12** 015705

Ekert A K 1991 Quantum cryptography based on Bell's theorem *Phys. Rev. Lett.* **67** 661–63

Huber D, Reindl M, Huo Y, Huang H, Wildmann J S, Schmidt O G, Rastelli A and Trotta R 2017 Highly indistinguishable and strongly entangled photons from symmetric GaAs quantum dots *Nat. Commun.* **8** 15506

Kaushal H and Kaddoum G 2017 Optical communications in space: challenges and mitigation techniques *IEEE Commun. Surv. Tutor.* **19** 57–96

Kwiat P G, Mattle K, Weinfurter H and Zeilenger A 1995 New high-intensity source of polarization-entangled photon pairs *Phys. Rev. Lett.* **75** 4337–41

Lambert S G and Casey W I 1995 *Laser Communications in Space* (Boston, MA: Artech House)

Little D J, Kitzler O, Abedi S, Alias A, Gilchrist A and Mildren R P 2021 Quantum-randomized polarization of laser pulses derived from zero-point diamond motion *Opt. Exp.* **29** 894–902

Makino K, Hashimoto Y, Yoshikawa J, Ohdan H, Toyama T, van Loock P and Furusawa A 2016 Synchronization of optical photons for quantum information processing *Sci. Adv.* **2** e1501772

Müller M, Bounouar S, Jöns K D and Michler P 2014 On-demand generation of indistinguishable polarization-entangled photon pairs *Nat. Photon.* **8** 224–28

NASA 2021 https://hubblesite.org/mission-and-telescope/the-telescope)

Pryce M H L and Ward J C 1947 Angular correlation effects with annihilation radiation *Nature* **160** 435

Salter C L, Stevenson R M, Farrer I, Nicoll C A, Ritchie D A and Shields A J 2010 An entangled-light-emitting diode *Nature* **465** 594–97

Shih Y H and Alley C O 1988 New type of Einstein–Podolsky–Rosen–Bohm experiment using pairs of light quanta produced by optical parametric down conversion *Phys. Rev. Lett.* **61** 2921–24

Singh A *et al* 2019 Quantum frequency conversion of a quantum dot single-photon source on a nanophotonic chip *Optica* **6** 563–69

Villar A *et al* 2020 Entanglement demonstration on board a nano-satellite *Optica* **7** 734–37

Ward J C 1949 *Some Properties of the Elementary Particles* (Oxford: Oxford University Press)

Yin Y *et al* 2012 Quantum teleportation and entanglement distribution over 100-kilometre free-space channels *Nature* **488** 185–88

IOP Publishing

Quantum Entanglement Engineering and Applications

F J Duarte and T S Taylor

Chapter 10

Quantum entanglement and the interpretations of quantum mechanics

The subject of interpretations of quantum mechanics is examined and discussed from a pragmatic perspective. The thoughts of Born, Dirac, Dyson, Feynman, Lamb, and Ward are applied to further develop a direct pragmatic paradox-free perspective of quantum mechanics. Particular attention is given to Dyson's *first layer* and *second layer* description of quantum mechanics. This perspective enables the description of quantum interferometry measurements and quantum entanglement measurements in a cohesive indeterministic approach via the concept of *quantum superposition probabilities*.

10.1 Introduction

Since the early days of quantum mechanics its interpretation has been the subject of controversy. Much of this controversy has been derived from attempts to reconcile the mathematical methodology of quantum mechanics with existing preconceived philosophical frameworks. Particular examples of these controversies are the incompatibility of quantum mechanics with possible hidden variable theories (von Neumann 1932) and arguments on the 'incompleteness' of quantum mechanics (Einstein *et al* 1935).

Quantum entanglement is certainly not exempt from this controversy. In fact, quantum entanglement is perhaps the phenomenon that most highlights the differences between classical and quantum thoughts. Philosophical arguments versus pragmatic mathematics. Indeed, quantum entanglement evolved in two parallel tracks. One philosophical and one physical (Duarte 2012, 2019). On the philosophical path, luminaries such as Einstein and Schrödinger criticized the newly found theory, while on the physical path physicists such as Dirac, Wheeler, Pryce, and Ward focused on the physics and the equations while practicing total silence on philosophical issues.

doi:10.1088/978-0-7503-3407-5ch10

Einstein, Podolsky, and Rosen (EPR) consider two separate systems while assigning the quantities P and Q to the second system. Their conclusion is that P and Q depend on the measurements performed in the first system which is decoupled from the second system. Hence they conclude: 'No reasonable description of reality could be expected to permit this' (Einstein *et al* 1935).

Schrödinger, who introduced the term *entanglement* to quantum mechanics, pondered about the issues raised by EPR. In this regard, he uses the word 'disconcerting' to describe the fact that there is no independence between systems that were initially entangled and that are no longer in physical proximity: 'after reestablishing one ψ-function… the other one can be inferred simultaneously' (Schrödinger 1935).

Bohr (1935) responded to the criticisms of EPR utilizing the principle of *complementarity* which is based on Heisenberg's uncertainty principle (Heisenberg 1927). However, his response was seven pages long, with very few equations, and was mainly based on a non-mathematical argument wading into philosophy. Bohr's rebuttal to EPR failed to quench the criticisms towards the orthodox interpretation of quantum mechanics.

EPR's argument is based on the premise that since the coordinate x has *no physical reality*, hence '*the quantum mechanical description of reality… is not complete*' (Einstein *et al* 1935). Although this argument is erroneous (Duarte 2014, 2019), it generated much of the energy behind the hidden variable theory programs (Bohm 1952a, 1952b, Bohm and Bub 1966), and the motivation to find alternative interpretations of quantum mechanics.

In this chapter issues of interpretation are considered and presented from a pragmatic perspective based in requirements for problem solving and experimental practice. In other words, the arguments are guided by the observation that *the ethos of physics… is the experiment.*

The material presented in this chapter is partly based on an extension of previous arguments and ideas discussed by Duarte (2014, 2019) whilst introducing several new reinforcing thoughts and insights.

10.2 Many alternative interpretations

Besides the original Copenhagen interpretation of quantum mechanics as developed by Bohr, Born, and Heisenberg there are at least a dozen alternative interpretations out there. And as pointed out by Dyson 'there are passionate arguments between believers [of the] various interpretations of quantum mechanics' (Dyson 2007). Here, some of the most prominent alternative interpretations are listed in alphabetical order: Bayesianism (QBism), decoherence, hidden variable theories, many-worlds interpretation, many-mind interpretation, objective collapse theories, stochastic interpretation, etc. To attempt description of any of these alternative interpretations is beyond the scope of this book.

From a pragmatic experimental perspective it should be stated that interest toward any alternative interpretation would have to be justified by a measurement, or an experiment, that could not be explained using the original Bohr–Born–Heisenberg interpretation. To date, no such experiments exist.

10.3 Guidance from quantum titans

In this section the opinions of Dirac, Feynman, Dyson, Lamb, and Ward, on interpretational issues in quantum mechanics are considered as guiding steps toward constructing an opinion on interpretations.

10.3.1 Dirac

Readers of this book should know that Paul Adrien Maurice Dirac, one of the fathers of quantum mechanics, was originally trained as an engineer (Dalitz 1987).

In his epic book *The Principles of Quantum Mechanics*, Dirac wrote about the interpretation of quantum mechanics but not in a manner as popularly perceived. These writings were about four pages long and were exclusively applied to further explain and elucidate his equations on quantum electrodynamics.

Despite his central role in the development of quantum theory, Dirac's statements on the broader issue of the interpretation of quantum mechanics are scarce. At a talk during a visit to Sydney, in 1975, he referred to the Einstein–Bohr debate via the three word sentence: 'Bohr was right' (Dirac 1975). Albeit he wisely left the door open to possible future developments.

More than a decade later he was explicit and direct on the interpretational issue: '*The interpretation of quantum mechanics has been dealt with by many authors, and I do not want to discuss it here. I want to deal with more fundamental things*' (Dirac 1987).

10.3.2 Feynman

Richard Feynman frequently referred to matters of interpretation. Some of the initial statements on this subject can be found in his famed volume III of *The Feynman Lectures on Physics* (Feynman *et al* 1965). Some of his thoughts are included here: '*the uncertainty principle protects quantum mechanics*' … '*unless a thing can be defined by measurement, it has no place in the theory*' … and he reminds the reader that 'indeterminacy' is also a feature of classical physics (Feynman *et al* 1965).

Feynman also commented on hidden variable theories whilst discussing interference.

In this regard he classified the existence of '*internal works—some inner variables*', as *impossible* (Feynman *et al* 1965). In further writings on this subject he stated '*the hidden variable theory… cannot be true*' (Feynman 1965).

Contrasting philosophy to science Feynman wrote: '*it is necessary for the very existence of science that minds exist which do not allow that nature must satisfy some pre-conceived conditions, like those of our philosopher*' (Feynman 1965). And from a broader vantage point he observed on the concept of uncertainty: 'All scientific knowledge is uncertain' (Feynman 1998).

Finally, in regard to the famous EPR paradox (Einstein *et al* 1935) Feynman appears to paraphrase Wheeler's thoughts, explaining that a polarization measurement on photon number one enables a prediction on the polarization of photon number two…

and he goes further: '*there doesn't seem to be any paradox at all… 'paradox' is only a conflict between reality and your feeling of what reality 'ought to be'* ' (Feynman *et al* 1965). It is interesting to note that these particular thoughts by Feynman have been continuously ignored in the literature.

One final observation: Feynman also applied quantum mechanical path integrals to describe *macroscopic* diffraction, or 'classical' diffraction, and thus led the way in applying quantum physics to macroscopic configurations (Feynman and Hibbs 1965).

10.3.3 Dyson

Another titan of quantum mechanics is Freeman Dyson. Two of his most iconic contributions are the proof of the equivalency of the quantum electrodynamics of Feynman, Schwinger, and Tomonaga (Dyson 1949a) and a paper that included a brilliant conjecture (Dyson 1949b) that inspired John Ward to discover his famed *Ward identities* (Ward 1950) which became of crucial importance to electromagnetic interactions and renormalization theory (Greiner and Reinhardt 2009, Close 2011).

In regard to quantum mechanics, Dyson wrote that in the world of quantum mechanics there are *two layers*: 'The first layer contains… the matrices of Heisenberg, and the state vectors of Dirac'. He adds that quantities belonging to the first layer can be accurately calculated but are not observable. Quantities belonging to the second layer include 'probabilities… intensities and polarizations… Quantities of the second layer can be directly observed but cannot be directly calculated' (Dyson 2007). He goes on to explain that measurables, such as intensity, are a function of the squares or products of first layer quantities. Furthermore, he suggests that nature is intrinsically linked to the mathematics of the first layer (Dyson 2007). Dyson's remarks highlight the extraordinary and crucial importance of probability amplitudes in the correct description of the physics. This topic is revisited in section 10.7.

10.3.4 Lamb

The *Lamb shift* measured in hydrogen (Lamb and Retherford 1947) was crucial to spur the developments in quantum electrodynamics disclosed by Feynman, Schwinger, and Tomonaga that were shown to be equivalent by Dyson (1949a).

Willis Lamb was openly critical of various ideas and concepts utilized to promote the need for alternative interpretations to the orthodox interpretation of quantum mechanics. In this regard he classified the EPR paradox as a 'misunderstanding' while referring to hidden variable theories and Bell's inequalities as 'unnecessary' (Lamb 2001).

In regard to quantum computing and quantum cryptography he wrote: 'The bulk of the papers on these subjects lack any understanding of the intrinsic probabilistic nature of quantum mechanics' (Lamb 2001). His criticism toward papers on teleportation was even more pronounced.

10.3.5 Ward

John Clive Ward did not say much about the interpretations of quantum mechanics and, just as Dirac, he was trained as an engineer. There were two exceptions via somewhat tenuous inferences in the late 1970s and early 1980s. First, in an interview with a student newspaper he said: 'Einstein did not accept quantum mechanics… he did not accept the statistical nature of the theory' (Ward 1979). Subsequently, around 1981, Ward openly advised one of us (FJD) to invest neither time nor effort in experiments aimed at testing hidden variable theories (Duarte 2011). Ward's contributions to the physics of the quantum era were classified as 'titanic' by Sakharov (1990) and are summarized and highlighted by several noted authors (Close 2011, Fraser 2008, Greiner and Reinhardt 2009, Kaiser 2005). More specifically, Freeman Dyson explains: 'the Ward papers on overlapping divergences demonstrated the deep connection between gauge invariance and renormalisability, which was another major step on the road to the standard model, namely the modern gauge theory of weak, electromagnetic and strong interactions' (as quoted by Dombey 2021).

Ward's thoughts on quantum mechanics are highly enlightening and here are reproduced with his own terse words: '*The inner mysteries of quantum mechanics require a willingness to extend one's mental processes into a strange world of phantom possibilities, endlessly branching into more and more abstruse chains of coupled logical networks, endlessly extending themselves forward and even backwards in time*' (Ward 2004).

His statement on quantum mechanics from the perspective of quantum entanglement is even more illuminating: '$(|x, y\rangle - |y, x\rangle)$ … *was my first lesson in quantum mechanics, and in a very real sense my last, since all the rest is mere technique, which can be learnt from books*' (Ward 2004).

10.4 Hidden variable theories

Hidden variable theories (HVT) have been mentioned in the literature since the early days of quantum mechanics. In this regard, John von Neumann refers to them explicitly in his book: 'the "hidden parameter" … has been proposed more than once' (von Neumann 1932). The HVT that caught the attention of von Neumann were based on 'dispersion free states,' that is, deterministic states. In this regard, the criticism of von Neumann towards HVT was direct and uncompromising: 'dispersion free ensembles, which would have to correspond to actual states… do not exist' (von Neumann 1932). Furthermore, in the same book, von Neumann offers a proof that concludes with the statement: 'It is therefore not, as is often assumed, a question of re-interpretation of quantum mechanics, the present system of quantum mechanics would have to be objectively false, in order that another description of the elementary processes than the statistical one be possible' (von Neumann 1932).

Despite von Neumann's firm opinion, a few years later Einstein and colleagues, in their famous EPR paper, questioned the completeness of quantum mechanics and in reference to a *complete* version of the theory they conclude 'we believe… that such a theory is possible' (Einstein *et al* 1935).

Encouraged by the EPR argument Bohm (1952a, 1952b) and Bohm and Bub (1966) provided hidden variable versions of quantum theory. Those theories are considered in the next section.

Moreover, John S Bell was not convinced by von Neumann's proof and thus openly criticized von Neumann's conclusion towards hidden variable theories. Bell produced his own proof on the incompatibility of quantum mechanics with hidden variable theories in what became known as the celebrated Bell's theorem or Bell's inequalities (Bell 1964, 1966). This theorem is considered here in subsection 10.4.2.

Also relevant to local realism is the statement: 'the two photons are entangled and according to local realism, their polarizations planes should become independent… a typical EPR situation. Already in 1948, observations agreed with quantum mechanics… not with local realism' (Dalitz and Duarte 2000).

10.4.1 Bohm's hidden variable theories

At least two hidden variable theories have been proposed in the open literature. The first one is due to Bohm (1952a, 1952b) and the second one is due to Bohm and Bub (1966). The hidden variable theory introduced by Bohm (1952a) concludes that 'the probability density of particles will cease to be equal to $|\Psi(x)|^2$' (Bohm 1952a). It also allows precise measurements of position and momentum, in violation of Heisenberg's uncertainty principle (Bohm 1952b).

Since Born's rule, $|\Psi(x)|^2 = \Psi(x)\Psi(x)^*$, is one of the central principles of quantum mechanics, it must be concluded that a theory that does not observe Born's rule *is not a quantum mechanical theory*. The same can be said about a theory that allows the violation of Heisenberg's uncertainty principle.

The second hidden variable theory considered here introduces an *irreversible* set of equations (Bohm and Bub 1966). This key antecedent leads to the conclusion that the Bohm and Bub hidden variable theory is also not compatible with the principles of quantum mechanics since *reversibility* is a key feature in the flow of quantum mechanical equations (Duarte 2019).

The observations advanced here simply state that these HVTs are not quantum mechanical. In this regard, it should have been recognized, as early as 1952, that the HVT proposed by Bohm was not quantum mechanical.

In 2006 a two-slit experiment using silicon droplets, rather than quanta or electrons, was found to produce interference patterns. The authors interpreted the result of this experiment as lending support to the 1952 Bohm's HVT (Couder and Fort 2006). However, a subsequent double-slit experiment also using droplets did not find evidence of interference thus casting 'strong doubt' on the Couder and Fort experiment and its interpretation (Andersen *et al* 2015).

10.4.2 Bell's theorem

Bell's theorem is a *classical statistical statement*. Its appeal rests on its mathematically straightforward approach to illustrate that hidden variable theories are incompatible with the results of quantum mechanics (Bell 1964). Bell reaffirmed von Neumann's earlier conclusion while claiming that von Neumann had reached such a conclusion using 'wanting' arguments (Bell 1964). Bell's criticism towards von Neumann's conclusion has been examined in detail by Bub (2010) whom concluded that von Neumann's proof does apply to the 'impossibility of recovering the quantum probabilities from an HVT of dispersion free (deterministic) states in which the quantum observables are represented as the 'beables' of the theory' According to Bell 'observables' must be *made* somehow out of beables' where 'beables' is made up from *be*-ables (Bell 1987).

There are several, but similar, approaches to present Bell's theorem. The exposition given here follows that outlined previously by Duarte (2014, 2019).

Bell defines a *hidden variable* λ, or *hidden variables* λ, as follows: 'It is a matter of indifference in the following whether λ denotes a single variable, or a set, or even a set of functions, and whether the variables are discrete or continuous' (Bell 1964).

In the development of Bell's theorem λ is defined as having a probability density $\rho(\lambda)$ such that its normalized probability is

$$\int \rho(\lambda) d\lambda = 1 \tag{10.1}$$

then the correlation between observables, which are a function of this hidden variable, $A(x, \lambda)$ and $B(y, \lambda)$ is stated by Bell as

$$P(x, y) = \int A(x, \lambda) B(y, \lambda) \rho(\lambda) d\lambda \tag{10.2}$$

x and y are 'setting directions' and the condition of *locality* requires that A depends only on x and B depends only on y. The variables x' and y' are assumed to be 'alternative settings' in the two measuring instruments. The experimenter determines these settings.

Furthermore, the following assumptions are made (Bell 1964)

$$A(x, \lambda) = \pm 1 \tag{10.3}$$

$$B(y, \lambda) = \pm 1 \tag{10.4}$$

$$|A(x, \lambda)| \leqslant 1 \tag{10.5}$$

$$|B(y, \lambda)| \leqslant 1 \tag{10.6}$$

Bell provides at least three approaches to derive his theorem (Bell 1964, 1966, 1971).

The approach exposed here begins by following Bell but then it is influenced by the discussion of Mandel and Wolf (1995). Albeit slightly different than previously disclosed derivations, the approach described here has the advantage of added transparency.

Following the format established in equation (10.2)

$$|P(x, y) - P(x, y')| \leqslant \int |[A(x, \lambda) B(y, \lambda) - A(x, \lambda) B(y', \lambda)]| \rho(\lambda) d\lambda \tag{10.7}$$

and

$$|P(x', y') + P(x', y)| \leqslant \int |[A(x', \lambda)B(y', \lambda) + A(x', \lambda)B(y, \lambda)]|\rho(\lambda)d\lambda \quad (10.8)$$

Adding equations (10.7) and (10.8)

$$\begin{aligned}|P(x, y) - P(x, y')| + |P(x', y') + P(x', y)| \leqslant \int |A(x, \lambda)[B(y, \lambda) - B(y', \lambda)]| \\ + |A(x', \lambda)[B(y', \lambda) + B(y, \lambda)]|\rho(\lambda)d\lambda\end{aligned} \quad (10.9)$$

Now, the RHS of this inequality includes *four distinct probabilities* as does the LHS. Thus the mathematical expression is balanced. Since the probabilities at hand obey the definitions $A(x, \lambda) = \pm 1$, $A(x', \lambda) = \pm 1$, $B(y, \lambda) = \pm 1$, $B(y', \lambda) = \pm 1$, then

$$\begin{aligned}\int |A(x, \lambda)[B(y, \lambda) - B(y', \lambda)]| + |A(x', \lambda)[B(y', \lambda) + B(y, \lambda)|]\rho(\lambda)d\lambda \\ = 2\end{aligned} \quad (10.10)$$

Substituting the result of equation (10.10) for the RHS of inequality (10.9) yields

$$|P(x, y) - P(x, y')| + |P(x', y') + P(x', y)| \leqslant 2 \quad (10.11)$$

as required by the central result of Bell's theorem (Bell 1964, 1966).

Substitution of quantum mechanical probabilities, from polarization quantum entanglement measurements, that is $P(x, y) \to P(\varphi_1, \varphi_2)$, $P(x, y') \to P(\varphi_1, \varphi_2')$, $P(x', y') \to P(\varphi_1', \varphi_2')$, and $P(x', y) \to P(\varphi_1', \varphi_2)$, into inequality (10.11) leads to

$$|P(\varphi_1, \varphi_2) - P(\varphi_1, \varphi_2')| + |P(\varphi_1', \varphi_2') + P(\varphi_1', \varphi_2)| \geqslant 2 \quad (10.12)$$

thus violating Bell's inequality.

As an example, consider the angular settings of $\varphi_1 = \pi/30$, $\varphi_2 = \pi/3$, $\varphi_1' = \pi/4$, $\varphi_2' = \pi/12$, in the superposition quantum probability (Pryce and Ward 1947)

$$|\psi\rangle|\psi\rangle^* = P(\varphi_1, \varphi_2) = -\cos 2(\varphi_1 - \varphi_2) \quad (10.13)$$

Calculating $P(\varphi_1, \varphi_2)$, $P(\varphi_1, \varphi_2')$, $P(\varphi_1', \varphi_2')$, and $P(\varphi_1', \varphi_2)$ and substituting those probabilities into

$$\Sigma_{|\psi\rangle|\psi\rangle^*} = |P(\varphi_1, \varphi_2) - P(\varphi_1, \varphi_2')| + |P(\varphi_1', \varphi_2') + P(\varphi_1', \varphi_2)| \quad (10.14)$$

yield

$$\Sigma_{|\psi\rangle|\psi\rangle^*} = |0.31 + 0.95| + |-0.50 - 0.87| = 2.63 \quad (10.15)$$

so that

$$\Sigma_{|\psi\rangle|\psi\rangle^*} \geqslant 2 \quad (10.16)$$

clearly violating Bell's inequality. On a finer point, it remains to be shown that classical probabilities of the type incorporated in Bell's inequality (10.11) are of the same type or class as superposition quantum probabilities such as $|\psi\rangle|\psi\rangle^*$.

The importance of Bell's work is that it provided a clear mathematical avenue to show that hidden variables and quantum mechanics were incompatible in a manner that was accepted by the practitioners in the field and by those interested in the philosophical aspects of quantum mechanics. However, that same conclusion can also be independently reached by observing the non-quantum mechanical features of Bohm's theories (see subsection 10.4.1).

Bell's theorem assumes the existence of hidden variable theories, also described as '*internal works*' by Feynman *et al* (1965), and its main application today is in the Ekert (1991) protocol as a tool to verify the non-intrusion of hidden third parties in quantum key distribution (see chapter 8).

Finally, it should be re-emphasized that the quantum feature in Bell's theorem is not its derivation but the ad hoc use of the superposition probability amplitude $|\psi\rangle = (|x\rangle_1 |y\rangle_2 - |y\rangle_1 |x\rangle_2)$ in calculating the superposition quantum probabilities alternatives that go into the inequality

$$|P(\varphi_1, \varphi_2) - P(\varphi_1, \varphi_2')| + |P(\varphi_1', \varphi_2') + P(\varphi_1', \varphi_2)| \leqslant 2$$

Note: in chapter 8 a simple all-quantum measurement protocol for quantum cryptography is introduced that uses the straight forward sum of superposition probabilities

$$\Sigma_{|\psi\rangle_- |\psi\rangle_-^*} = |P(\varphi_1, \varphi_2)| + |P(\varphi_1, \varphi_2')| + |P(\varphi_1', \varphi_2')| + |P(\varphi_1', \varphi_2)| \tag{10.17}$$

instead of Bell's theorem equation (10.14). In this $\Sigma_{|\psi\rangle_- |\psi\rangle_-^*}$ the four individual absolute values of the superposition probabilities, corresponding to the angular settings (φ_1, φ_2), (φ_1, φ_2'), (φ_1', φ_2'), (φ_1', φ_2), are added separately.

10.4.3 Comment on the EPR paper

Although the existence of hidden variable theories was alluded to by von Neumann (1932), in his book *Mathematische Grundlagen der Quanten-Mechanik*, it was Einstein *et al* (1935) that ignited the debate on the possible 'incompleteness' of quantum mechanics. In that paper Einstein suggested that a 'complete' version of quantum mechanics was *possible* (Einstein *et al* 1935). Indeed, it was the EPR paper that encouraged Bohm and others to pursue hidden variable theories (Bohm 1952a, Bohm and Bub 1966).

As suggested by Duarte (2014), Bohr could have neutralized Einstein's argument had he focused his response just on a strict measurements discussion involving $\Delta p\, \Delta x \approx h$.

In his response Bohr did mention the uncertainty principle but also engaged in a rather long and broad seven page somewhat philosophical discussion that left part of the readership wondering.

Heisenberg's uncertainty principle $\Delta p\, \Delta x \approx h$ indicates that a measured non-zero value for Δp immediately leads to

$$\Delta x \approx \frac{h}{\Delta p} \tag{10.18}$$

which is a *finite* number. That is the physics. However, the initial part of the EPR argument rests on the premise of an 'all values' spread in the coordinate x (Einstein *et al* 1935). This all-values spread led EPR to ascertain: '*when the momentum of the particle is known its coordinate has no physical reality*' (Einstein *et al* 1935). However, as pointed out by Duarte (2014, 2019), *no physical reality* for x can only occur if the momentum $p \pm \Delta p$ is measured *exactly* as p with $\Delta p = 0$, which is physically impossible. As articulated by one of the great masters: '*the uncertainty principle protects quantum mechanics*' (Feynman *et al* 1965). Furthermore, uncertainty in measurements has been part of physics since its genesis, see for example Newton (1686).

Allowing a logical progression based on the premise of an exact measurement with zero error is a situation that might only arise in the absence of experience with measurements and experimental physics. Classical or quantum.

10.5 A pragmatic perspective on the interpretations of quantum mechanics

Born wrote that the principles of quantum mechanics, that is, orthodox quantum mechanics, were '*found by a slow and tedious process of interpreting experimental results*' (Born 1949). In this effort he included Bohr, Heisenberg, and himself. In reference to an improved quantum mechanics he added: 'I think that the indeterministic foundations will be permanent… There remains now only to show how the ordinary, apparently deterministic laws of physics can be obtained from these foundations' (Born 1949). This is a profound statement continuously ignored in the contemporaneous literature.

The standard interpretation of quantum mechanics is also known as the *Copenhagen interpretation* and the *orthodox interpretation*. It is the interpretation initially developed by Bohr, Born, and Heisenberg. In its most basic form it adheres to the principle of complementarity, which is related to Heisenberg's uncertainty principle (Bohr 1935), the probability amplitude to describe the physics (Dirac 1958), and Born's rule to calculate probabilities (Born 1926). Probability amplitudes and wave functions are homologous. Dirac added further insights in regard to interference and superposition (Dirac 1958).

Feynman, who implicitly accepted the standard interpretation… '*the uncertainty principle protects quantum mechanics*' (Feynman *et al* 1965)… did not refer to this interpretation in philosophical terms but rather stated a brief set of mathematical principles.

Philosophy can be steered by cultural biases. Therefore, it is advantageous to keep philosophical tendencies at a distance and to adopt an approach that is entirely guided by measurements and experimental results. Pragmatism brings us closer to nature. From a pragmatic perspective, which is the perspective of this book, the following observations are relevant (Duarte 2019):

1. A theory should be criticized if it fails to *reproduce* experimental results. This *is not* the case with quantum mechanics.
2. A theory should be criticized if it fails to *predict* experimental results. This *is not* the case with quantum mechanics.

3. A theory should be criticized if it fails to predict measurable phenomena. This *is not* the case with quantum mechanics.
4. A theory should even be criticized if it leads to messy equations including ad hoc parameters. This is neither the case with quantum interference nor quantum entanglement equations.

As suggested by Duarte (2019) an excellent manner to judge the validity of a theory is to apply Einstein's criterion: 'The correctness of the theory is judged by the degree of agreement between the conclusions of the theory and human experience. This experience… in physics takes the form of experiment and measurement' (Einstein *et al* 1935). Quantum mechanics amply meets this criterion.

It can be further argued that pragmatism, guided by experimental evidence, is the approach adopted toward the principles of quantum mechanics by luminaries such as Born, Dirac, Feynman, and Lamb. In this regard, Lamb goes further and suggests, as Born did earlier (Born 1949), that the probabilistic principles of quantum mechanics should be applied to describe macroscopic and classical phenomena (Lamb 1987b). In this regard, it can indeed be shown how, in a hierarchic order, crucial optics can be explained starting from quantum interference → diffraction → refraction → reflection (Duarte 1997, 2014). A whole chapter is devoted to this topic in *Fundamentals of Quantum Entanglement* (Duarte 2019).

Previously it has been suggested that the best interpretation of quantum mechanics is… '*no interpretation at all*' (Duarte 2014). What that means is that, from a pragmatic perspective, an experimentalist perspective, or an engineering perspective, the best approach is to use, apply, and enjoy the proven principles of quantum mechanics since these principles are provided by nature itself via experimental evidence.

10.6 Quantum principles

From an interferometric perspective the very basic quantum mechanical framework needed to make the physics work can be based on the following laws, or principles:

1. Theorem: *The probability amplitude for a single quantum is mathematically identical to the probability amplitude for an ensemble of indistinguishable quanta.*
2. The probability amplitude for a single quantum, or an ensemble of N indistinguishable quanta, is given by the Dirac–Feynman principle (Feynman *et al* 1965)

$$\langle d|s\rangle = \sum_{i=1}^{N} \langle d|j\rangle\langle j|s\rangle \tag{10.19}$$

   The N in this superposition series means that every propagation path must be incorporated in the calculation.
3. Equation (10.19) is inherently linked to the *principle of superposition*: 'Any state may be considered as the result of a superposition of two or more

other states… conversely any two or more states may be superposed to give a new state' (Dirac 1958).

A further example of superposition is the interferometric probability amplitudes

$$|\psi\rangle_+ = 2^{-1/2}(|\alpha\rangle + |\beta\rangle) \tag{10.20}$$

$$|\psi\rangle_- = 2^{-1/2}(|\alpha\rangle - |\beta\rangle) \tag{10.21}$$

4. The probability amplitudes can be represented by complex wave functions of the form (Dirac 1958)

$$\Psi = \Psi_0 e^{-i(\omega t - kx)} \tag{10.22}$$

or more specifically

$$\langle j|s\rangle = \Psi(r_{j\leftarrow s})e^{-i\theta} \tag{10.23}$$

$$\langle d|j\rangle = \Psi(r_{d\leftarrow j})e^{-i\varphi} \tag{10.24}$$

5. The physics of quantum mechanics is *reversible*. For instance, the probability amplitude for quantum entanglement can be derived from the Dirac–Feynman principle (Duarte 2019)

$$\langle d|s\rangle = \sum_{j=1}^{N}\langle d|j\rangle\langle j|s\rangle \rightarrow |\psi\rangle = 2^{-1/2}(|x\rangle_1\,|y\rangle_2 \pm |y\rangle_1\,|x\rangle_2) \tag{10.25}$$

or *vice versa* (Duarte 2019)

$$|\psi\rangle = 2^{-1/2}(|x\rangle_1\,|y\rangle_2 \pm |y\rangle_1\,|x\rangle_2) \rightarrow \langle d|s\rangle = \sum_{j=1}^{N}\langle d|j\rangle\langle j|s\rangle \tag{10.26}$$

6. Additional Dirac principles as outlined by Feynman *et al* (1965)

$$\langle j|i\rangle = \delta_{ji} \tag{10.27}$$

$$\langle d|s\rangle = \langle s|d\rangle^* \tag{10.28}$$

7. Dirac's identity for indistinguishable quanta in different states (Dirac 1958)

$$|X\rangle = |a\rangle_1\,|b\rangle_2\,|c\rangle_3 \dots |g\rangle_n \tag{10.29}$$

8. Born's rule: the probability is given by multiplying the probability amplitude with its complex conjugate (Born 1926)

$$|\langle d|s\rangle|^2 = \langle d|s\rangle\langle d|s\rangle^* \tag{10.30}$$

Probability amplitudes such as $\langle d|s\rangle$ are complex numbers and thus have neither magnitude nor absolute value. The straightforward multiplication of the probability amplitude with its complex conjugate via Born's rule $\langle d|s\rangle\langle d|s\rangle^*$ provides a probability less than or equal to one.

9. The measured intensity, at a given frequency, is proportional to the quantum probability

$$I \propto \langle d|s\rangle\langle d|s\rangle^* \tag{10.31}$$

10. 'Photons cannot be localized… and they do not behave at all like particles' (Lamb 1995). This is totally compatible with Heisenberg's uncertainty principle.
11. Heisenberg's uncertainty principle (Heisenberg 1927)

$$\Delta p \, \Delta x \approx h \tag{10.32}$$

This principle provides the foundations for the *indeterminacy* that is one of the crucial characteristic of quantum mechanics.

12. Planck's quantum energy equation (Planck 1901) which gave birth to quantum mechanics

$$E = h\nu \tag{10.33}$$

These laws, and/or principles, of quantum mechanics provide the foundations of the knowledge needed to quantify generalized quantum interference and quantum entanglement.

10.7 Quantum measurements

One of the areas in quantum mechanics that continues to generate questions and uncertainties is the area of quantum measurements. This has become known as the 'quantum measurement problem.' This is despite attempts by well-known quantum physicists such as Landau and Lifshitz (1977), Lamb (1986a, 1986b, 2001), and van Kampen (1988), to clarify the physics of quantum measurements. In this regard, Lamb observes: 'None of the physicists who founded quantum mechanics… explained how measurements of physical quantities could be made' (Lamb 1987a). Strictly speaking, however, it can be argued quite effectively that since quantum mechanics accurately reproduces and predicts the results of experimental measurements there is no such problem as the measurement problem. In this regard, 'the measurement problem' is not a practical tangible problem but an elusive conceptual problem.

Here, a description of quantum measurements is provided from the perspective of probability amplitudes, specifically interferometric and quantum entanglement probability amplitudes. This is done utilizing Dyson's classification of *first layer* and *second layer* quantum mechanics.

10.7.1 Dyson's first layer quantum mechanics

Dyson's *first layer* is the mathematical level at which nature works. This layer of nature is dominated by the probability amplitudes that correctly describe the physical configuration of the experiment.

Considering a photon emitted at a source s, propagating via a transmission grating j, and arriving to the interferometric plane d, the physics is accurately described by the Dirac–Feynman interferometric principle which is the superposition probability amplitude

$$\langle d|s\rangle = \sum_{j=1}^{N} \langle d|j\rangle\langle j|s\rangle$$

Here, it should be noted that the interferometric plane d is also the detector's plane. If the physics is modified by the experimenter via the insertion of an additional intermediate plane such as a beam splitter k, introduced between j and d as illustrated in figure 10.1, then the probability amplitude undergoes a severe, and *immediate*, alteration becoming

$$\langle d|s\rangle = \sum_{k=1}^{N}\sum_{j=1}^{N} \langle d|k\rangle\langle k|j\rangle\langle j|s\rangle \tag{10.34}$$

This probability amplitude introduces three intra interferometric propagation distances $D_{\langle j|s\rangle}$, $D_{\langle k|j\rangle}$, and $D_{\langle d|k\rangle}$ rather than the originals $D_{\langle j|s\rangle}$ and $D_{\langle d|j\rangle}$. The beam splitter allows *information to be extracted* from the propagating interferogram. In this situation the original propagating interferogram described by equation (10.19) ceases to exist and immediately a new severely altered interferogram, described by equation (10.34), is observed (Duarte 2002). Even if the beam splitter is replaced by a single strand of a transparent microscopic fiber, such as a spider web silk fiber, as illustrated in figure 10.2, the probability amplitude is transformed to (Duarte *et al* 2013, Duarte 2019)

$$\langle d|s\rangle = \sum_{j=1}^{N} \langle d|k\rangle\langle k|j\rangle\langle j|s\rangle \tag{10.35}$$

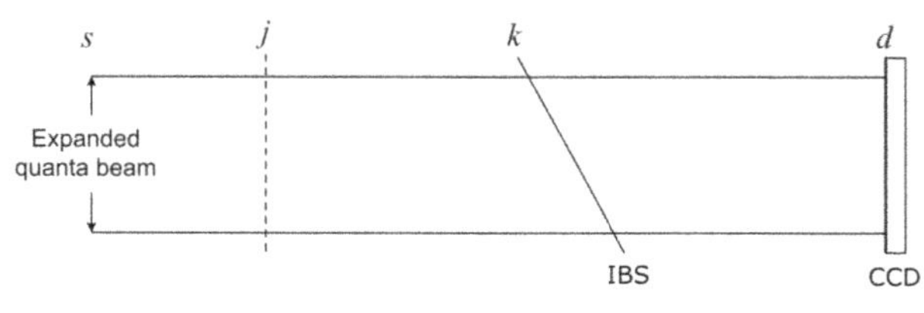

Figure 10.1. N-slit interferometer modified by an intra interferometric beam splitter (IBS). The IBS allows information, from a distorted interferogram, to be extracted (Duarte 2002).

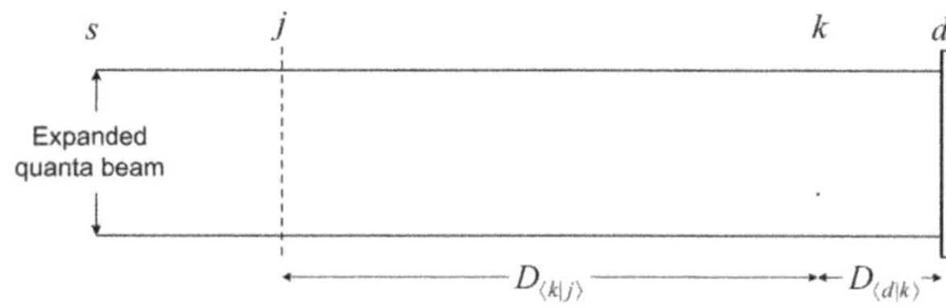

Figure 10.2. N-slit interferometer modified by a transparent spider web silk fiber. The subtle distortion induced by the transparent microscopic fiber on the propagating interferogram can be accurately predicted, or reproduced, via the interferometric principles described in the text (Duarte *et al* 2013).

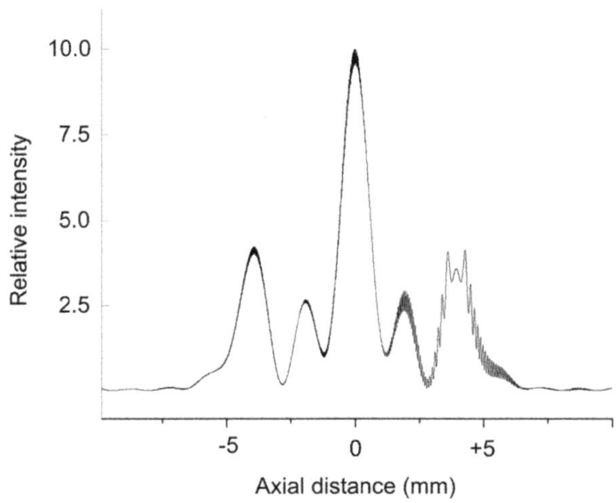

Figure 10.3. Theoretical interferogram corresponding to $N = 3$ being intercepted by a transparent spider web silk fiber with a 25 μm diameter at $D_{\langle k|j\rangle} = 7.235 - 0.15$ m, or $D_{\langle d|k\rangle} = 0.15$ m. In this calculation $w = 50$ mm, $w_s = 50$ mm, and $\lambda = 632.82$ nm (Duarte *et al* 2013). Reprinted by permission of Taylor & Francis.

Certainly, these transformations to the probability amplitude immediately alter the corresponding probability and ultimately the measured spatial intensity profile that depends directly on the probability distribution. It should be emphasized that the transition from equations (10.19) to (10.34) or (10.35) is instantaneous and could be classified as a *jump* in probability amplitudes or a jump in superposition states. The predicted interferogram for the experimental configuration depicted in figure 10.2 is shown in figure 10.3 while the corresponding measured interferogram is shown in figure 10.4 (Duarte *et al* 2013).

Unimpeded propagation and permanence of the interferometric plane *exactly perpendicular* to the propagation axis, and to the expanded interferometric distribution, precisely at *d*, *so that no information can be extracted by an external observer*, leaves the original probability amplitude distribution

$$\langle d|s\rangle = \sum_{j=1}^{N} \langle d|j\rangle\langle j|s\rangle$$

undisturbed.

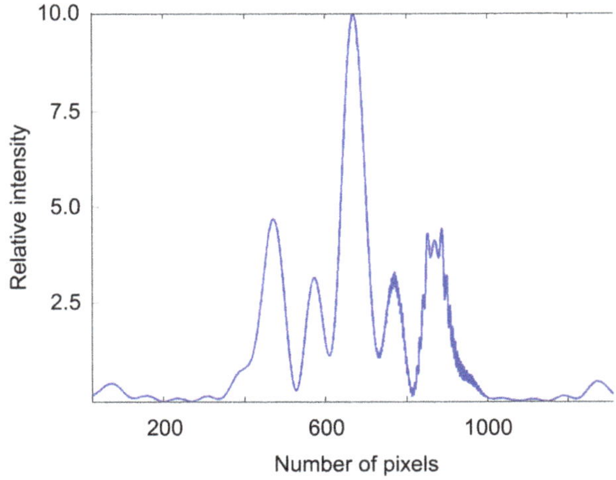

Figure 10.4. Measured interferogram corresponding to $N = 3$ being intercepted by a transparent spider web silk fiber with a 25 μm diameter at $D_{\langle k|j\rangle} = 7.235 - 0.15$ m, or $D_{\langle d|k\rangle} = 0.15$ m. In these measurements $w = 50$ mm, $w_S = 50$ mm, and $\lambda = 632.82$ nm (Duarte *et al* 2013). Reprinted by permission of Taylor & Francis.

The probability amplitude $\langle d|s\rangle$ is a superposition of base states. More specifically, a mathematical series of the form

$$\langle d|s\rangle = \langle d|1\rangle\langle 1|s\rangle + \langle d|2\rangle\langle 2|s\rangle + \langle d|3\rangle\langle 3|s\rangle \ldots + \langle d|N\rangle\langle N|s\rangle \tag{10.36}$$

If a single photon illuminates the whole grating of N slits (Duarte 2003), and if this photon undergoing interference is left undisturbed, it propagates in this $\langle d|s\rangle$ superposition until a measurement is performed. Prediction of the correct experimental result depends on considering the entire series of the probability amplitude superposition $\langle d|s\rangle$ in order to calculate the superposition probability. This is an observation originally due to Feynman (Feynman and Hibbs 1965).

If the initial condition measurement parameters N, slit width w, slit separation w_S, and wavelength λ are determined with priority then it is the interferometric distance $D_{\langle d|j\rangle}$ that continues to evolve. At the instant of the measurement, at the final $D_{\langle d|j\rangle}$, all the base states acquire their final numerical value or 'weight' thus yielding the final superposition probability amplitude $\langle d|s\rangle_{D_{\langle d|j\rangle}}$. This description of single-photon interference is entirely compatible with the non-local nature of the photon as derived from Heisenberg's uncertainty principle.

In the case of $n = N = 2$ quantum entanglement the probability amplitude that describes the physics is

$$|\psi\rangle = 2^{-1/2}(|x\rangle_1\,|y\rangle_2 - |y\rangle_1\,|x\rangle_2) \tag{10.37}$$

Again, this superposition evolves as the entangled quanta propagate and the base states acquire their final setting as the measurement is performed. From Dyson's perspective the superposition probability amplitude $|\psi\rangle$ becomes the superposition probability $|\psi\rangle|\psi\rangle^* = P(\varphi_1, \varphi_2)$ and *it is this superposition probability* that determines whether the measurer measures a photon with x polarization or a photon with y polarization. This interpretation is entirely compatible with Dirac's discussion on superposition: whilst considering the superposition of two states, A and B, Dirac emphasized that the result of a measurement on the superposed state is determined by the probability 'depending on the relative weights of A and B in the superposition' (Dirac 1958).

In both cases, quantum interference and quantum entanglement, the final weight of the states comprising the superposition is entirely decided by the indeterministic, or statistical, nature of quantum mechanics. Further, it is the whole superposition, the superposition mathematically configured by all the states, that determines the correct mathematical configuration of the probability (Duarte 1997, 2014).

10.7.2 Dyson's second layer quantum mechanics

Dyson's second layer is the physical realm of the measurables. This layer is all about probabilities and physical quantities such as intensities and polarizations (Dyson 2007).

Dyson's second layer is where the superposition probability amplitude described by equations (10.19) or (10.35) becomes a quantum probability via Born's rule $\langle d|s\rangle\langle d|s\rangle^*$.

The quantum intensity corresponding to this probability is (see chapter 2)

$$\mathscr{I}(\nu) = \mathrm{K}\, h\nu\langle d|s\rangle\langle d|s\rangle^* \tag{10.38}$$

or more explicitly

$$\mathscr{I}(\nu) = \mathrm{K}\, h\nu\sum_{j=1}^{N}\Psi(r_j)\sum_{m=1}^{N}\Psi(r_m)e^{i(\Omega_m-\Omega_j)} \tag{10.39}$$

and this is the interferometric distribution that is observed at the interferometric plane or detection surface d, be it a CCD array, a CMOS array, a photographic plate, or a collection of rods in a human retina. This means that we are not dealing just with an interferometric probability anymore but with a physical *measurable intensity* which, for a given ν, is proportional to the quantum probability, that is $\mathscr{I}(\nu) \propto \langle d|s\rangle\langle d|s\rangle^*$. In this description the probability amplitude $\langle d|s\rangle$ is at all times associated to the probability $\langle d|s\rangle\langle d|s\rangle^*$ and therefore to its corresponding intensity distribution $\mathscr{I}(\nu) = \mathrm{K}\, h\nu\langle d|s\rangle\langle d|s\rangle^*$.

The single photon distributed throughout the $\langle d|s\rangle\langle d|s\rangle^*$ spatial profile may or may not trigger the development of a single charge at a particular site on the surface of the spatial detector. If it does, the phenomenon manifests as a single dot in a two-dimensional surface of un-triggered pixels. In other words, a single charge is created within the allowed probabilistic $\langle d|s\rangle\langle d|s\rangle^*$ spatial profile and additional charges, corresponding to additional pixels, are created with the arrival of additional quanta. Note that the 'single dot' observed at the detection screen is just a local surface charge created by the interaction of the detection surface and the $\mathscr{I}(\nu) = \mathrm{K}\, h\nu\langle d|s\rangle\langle d|s\rangle^*$ distribution. From this perspective, the sudden appearance of a single charge dot does not mean that the photon has become 'localized.' For the cumulative detection of single photons, the process described above continues to occur randomly within the $\langle d|s\rangle\langle d|s\rangle^*$ spatial distribution until the entire interferometric profile is revealed (Duarte 2004).

The measuring process for a population of indistinguishable photons, or an ensemble of r indistinguishable photons, is along the same lines except that the intensity this time is given by

$$\mathscr{I}(\nu)_r = r\mathrm{K}\, h\nu\langle d|s\rangle\langle d|s\rangle^* \tag{10.40}$$

$$\mathscr{I}(\nu)_r = r\mathrm{K}\, h\nu\sum_{j=1}^{N}\Psi(r_j)\sum_{m=1}^{N}\Psi(r_m)e^{i(\Omega_m-\Omega_j)} \tag{10.41}$$

where r is the number of indistinguishable quanta. In this case the indistinguishable quanta remain within the $|\langle d|s\rangle\langle d|s\rangle^*|$ distribution and collectively, at once, overcome the threshold of the photographic film or the spatial digital detector. In this manner the registered pattern is a faithful image of the $\langle d|s\rangle\langle d|s\rangle^*$ distribution. Experimentally, narrow-linewidth laser intensity interferometric profiles expressed as a function of $|\langle d|s\rangle\langle d|s\rangle^*|$ have been accurately confirmed in the $2 \leqslant N \leqslant 2000$ range (Duarte 1993, 2003).

Future ultra high-sensitivity spatial detectors should be capable of mapping probability distributions rather than just localized charges generated via the photon–matter interaction. That is, probability distribution detectors for single-photon radiation.

10.7.2.1 Time flow

In his description of quantum mechanics Ward referred to calculations '*... endlessly extending themselves forward and even backwards in time*' (Ward 2004). Up to now the subject of time (t), which is intimately related to entropy, has not surfaced in our discussions since the parameter t is absent in the Diracian probability amplitudes. As will be shown next, time manifests itself in Dyson's second layer via the probabilistic intensity distributions, i.e. equation (10.41), which are a function of the superposition quantum probability $\langle d|s\rangle\langle d|s\rangle^*$.

Consider a transmission grating with $N = 30$ illuminated at $\lambda = 593.93$ nm, the slit width is $w = 50$ μm and each slit is separated by $w_s = 50$ μm. For an intra-interferometric distance of $D_{\langle d|j\rangle} = 0.03$ m the near-field, highly structurized, interferometric distribution is illustrated in figure 10.5. However, in the far field, at a time $t = t_i + \Delta t$, this pattern looses its distinct spatial structure and tends to assume the smooth profile of a near-Gaussian distribution, as illustrated in figure 10.6 for $D_{\langle d|j\rangle} = 30.00$ m. Assuming $c = 2.997\,924\,58 \times 10^8$ ms then setting $t_i = 0$ means that, at the intra-interferometric distance of $D_{\langle d|j\rangle} = 30.00$ m, $t_i + \Delta t = 100.069\,228\,56$ ns and we have an interferometric clock! Smooth near-Gaussian distributions are typical of single-transverse-mode single-longitudinal-mode (SLM) laser emission (Duarte 2003).

The highly structurized interferometric distribution in the near field, at t_i, is indicative of low entropy while the smooth near-Gaussian higher-divergence

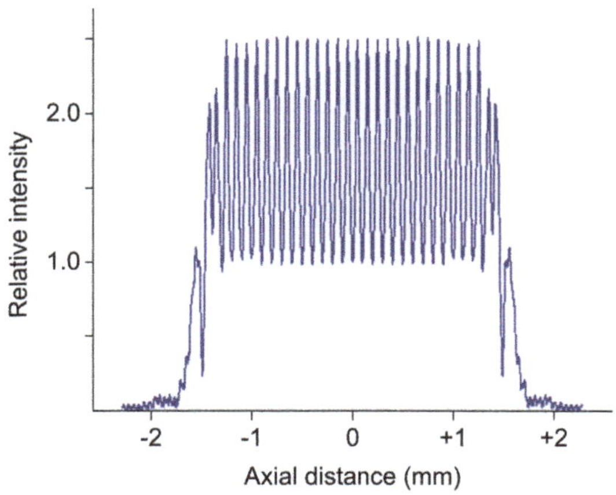

Figure 10.5. Calculated interferograms corresponding to $N = 30$ for an intra-interferometric distance of $D_{\langle d|j \rangle} = 0.03$ m. This is a low entropy regime (see text). In this calculation $w = 50$ μm, $w_s = 50$ μm, and $\lambda = 593.93$ nm.

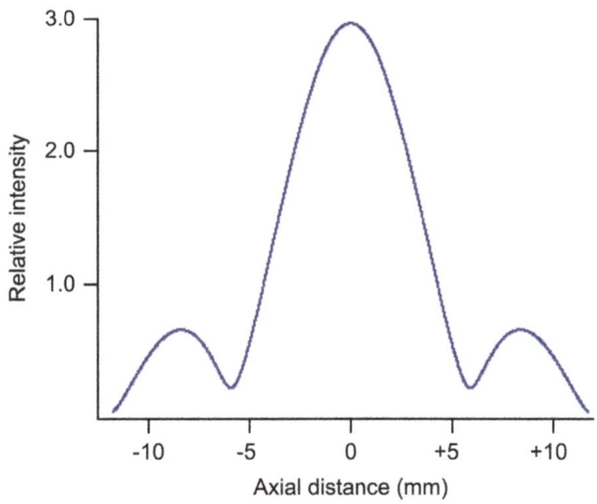

Figure 10.6. Calculated interferograms corresponding to $N = 30$ for intra-interferometric distances of $D_{\langle d|j \rangle} = 30$ m. This is a higher entropy regime (see text). In this calculation $w = 50$ mm, $w_s = 50$ mm, and $\lambda = 593.93$ nm.

distribution in the far field, at $t = t_i + \Delta t$, is indicative of a higher regime of entropy. Thus, it can be concluded that these interferometric distributions do include intrinsic information that can be used to characterize the degree of entropy, or the *arrow of time*, associated with the propagation of quanta.

Here, it should be indicated that the broad near-Gaussian beam illustrated in figure 10.6, for $D_{\langle d|j \rangle} = 30.00$ m, can be collected and *beam compressed* (Duarte 2016a) to approximately reproduce the initial illumination field and again illuminate the $N = 30$ slit array and approximately reproduce the propagation physics.

Thus we can go from low interferometric entropy to high interferometric entropy and create a series of high-to-low to high-to-low... entropy regimes. The word *approximately* is used since losses in the beam shaping optics will reduce slightly the illumination intensity each time a beam transition is executed. Here, we should mention that the concept of entropy reduction appears to have been first introduced by Schrödinger in his wonderful book *What is life?* where he links life with negative entropy (Schrödinger 1944), see also chapter 8.

One more aspect of these probabilistic superposition calculations is that once the experimenter decides and sets the initial conditions, that is λ, N, slit width (w), and slit separation (w_s), equation (10.41) predicts the measured interferometric distribution for any given intra interferometric distance $D_{\langle d|j\rangle}$ in the near and far field. That is, these measurements are completely predictable by the superposition interferometric probability once the initial conditions are set by the measurer.

10.7.3 Quantum measurement sequence

In addition to the ideas presented so far, the following description is relevant: a photon originates from a definite position in space as emitted by a single-photon source or a single-transverse-mode SLM source s. Next, it is expanded orthogonal to the propagation axis and made to interfere via an N-slit array, j. The photon undergoing interference then propagates, within the $|\langle d|s\rangle\langle d|s\rangle^*|$ interferometric distribution in free space, until it arrives at the interferometric plane d. The overall sequence of events can be expressed as (Duarte 2004)

$$|s\rangle \rightarrow \langle d|j\rangle\langle j|s\rangle \rightarrow \langle d|s\rangle\langle d|s\rangle^* \rightarrow \langle D|i\rangle\langle i|d\rangle \rightarrow \langle D|d\rangle\langle D|d\rangle^* \tag{10.42}$$

The crucial step

$$\langle d|s\rangle\langle d|s\rangle^* \rightarrow \langle D|i\rangle\langle i|d\rangle \tag{10.43}$$

is where the original interferometric distribution characterized by $\langle d|s\rangle\langle d|s\rangle^*$ is transferred to the classical detector. It is here where photonic energy distributed in the $\mathscr{I}(\nu)$ profile enters the photo-electric domain of the detector and creates electron charges within the $\mathscr{I}(\nu) = \mathrm{K}\, h\nu|\langle d|s\rangle\langle d|s\rangle^*|$ spatial profile. This is assumed to be a destructive non-reversible transition where the interferometric quantum distribution $\mathscr{I}(\nu)$ is spatially registered for further amplification.

An alternative description of the measurement mechanism is to think, in Dyson's first layer, of a constantly evolving superposition of base states $\langle d|s\rangle$, as the interfering quanta propagates, with each of the combined base states $\langle d|j\rangle\langle j|s\rangle$ themselves also evolving. At the instant of the measurement the superposition $\langle d|s\rangle$ becomes $\langle d|s\rangle\langle d|s\rangle^*$ with each base state in the series becoming fixed at its measurement mathematical value. The measurement itself resides in Dyson's second layer.

The transition

$$\langle D|i\rangle\langle i|d\rangle \rightarrow \langle D|d\rangle\langle D|d\rangle^* \tag{10.44}$$

describes the process within the macroscopic detector from arrival at its surface d to a display screen D.

10.7.4 Quantum entanglement measurements

Returning to Dyson's second layer: as already mentioned, for the case of $n = N = 2$ quantum entanglement, as illustrated in figure 10.7, the superposition probability amplitude that describes the physics is

$$|\psi\rangle = 2^{-1/2}(|x\rangle_1\,|y\rangle_2 - |y\rangle_1\,|x\rangle_2)$$

If a measurement at the $+z$ side of the source yields a photon polarized in the x-direction then the quanta that propagated to the $-z$ side of the source is polarized in the y-direction. In quantum entanglement situations the superposition probability amplitude $|\psi\rangle$ ultimately reduces to either an x-polarization or a y-polarization with each reduction outcome occurring with a probability $P(\varphi_1, \varphi_2)$.

The superposition probability $P(\varphi_1, \varphi_2)$ is calculated by multiplying the superposition $|\psi\rangle$ with its complex conjugate, according to Born's rule, so that (Pryce and Ward 1947)

$$\begin{aligned} P(\varphi_1, \varphi_2) &= |\psi\rangle|\psi\rangle^* \\ P(\varphi_1, \varphi_2) &= -\cos 2(\varphi_1 - \varphi_2) \end{aligned} \tag{10.45}$$

It is important to emphasize that it is this superposition probability, which lies in the numerical range

$$0 \leqslant P(\varphi_1, \varphi_2) \leqslant 1 \tag{10.46}$$

that determines the outcome of the measurements.

10.7.5 The observer

Quantum measurements certainly are influenced by *the observer*, *the measurer*, or *the experimentalist*:

1. The measurer selects what to measure. For example, a quantum entanglement measurement or a quantum interferometric measurement.
2. The measurer selects the tool to be used in the measurement. In the case of quantum entanglement the measurer can use rotating polarizer analyzers or fixed detector configurations via polarizing beam splitters.

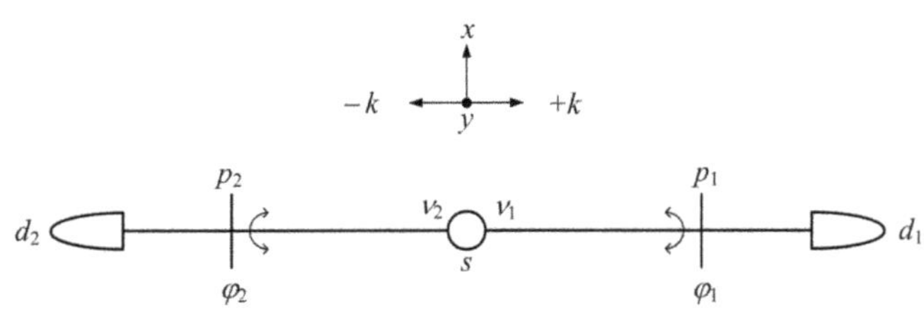

Figure 10.7. Generic quantum entanglement configuration in the visible optical domain for $n = N = 2$.

3. Once the measurement and the measuring instrument are selected, the measurer also plays a direct role in the process since the recorded probabilities will depend on the surface quality of the optics selected by the measurer, the electronic noise of the system built by the measurer, and the ability or skill of the measurer to optimally align the optics.

Hence, just as in the classical case, the measurer does affect significantly the uncertainty (i.e. errors) and reliability of the measurements.

However, the quantum situation is far more susceptible to the measurer's behavior since the results of the measurements can be severely altered, even catastrophically demolished, if the measurer, for instance, tries to intrude into a measurement in progress (Duarte 2002, 2014).

This probability amplitude description of quantum measurements involves 'jumps' in the probability amplitude when the propagation physics, via geometrical modifications, is altered.

In the case of quantum entanglement measurements the probability amplitude $|\psi\rangle$ ultimately allows either x or y polarizations to be measured with a superposition probability $P(\varphi_1, \varphi_2) = |\psi\rangle|\psi\rangle^*$. In this description the detection part of the observer can either be a human retina or a physical element such as a recording pixel screen or the surface of a silver-halide photographic film.

10.7.6 On 'the collapse of the wave function'

In this section, quantum measurements have been discussed without utilizing the concept of 'the collapse of the wave function.' In the literature this collapse of ψ can often be found associated with the existence of the 'quantum measurement problem' (Albert and Vaidman 1989, Namiki and Pascazio 1992, Penrose 1996). The collapse of ψ is described by Bohm in terms of a spatial collapse from a broad region to a narrow region (Bohm 1952a). In contemporaneous terms the collapse of ψ is often described as a sudden transition from a series of superposed eigenstates to a single eigenstate. For the benefit of this discussion, at this stage, it should be reiterated that wave functions and probability amplitudes are synonymous (Feynman *et al* 1965) and as such they exist in Dyson's first layer.

Two initial observations about the collapse of the wave function are rather useful: first, this concept is absent from the discussion on quantum superposition in Dirac's landmark book (Dirac 1958). Secondly, although Feynman dedicates two separate sections in his lectures to the discussion of wave functions, he neither refers to nor discusses 'the collapse of the wave function' (Feynman *et al* 1965).

In our framework, using N-slit interferometry as a vehicle, the concept of the collapse of the wave function would have to be described in terms of a sudden reduction of the superposition probability amplitude $\langle d|s\rangle$ to just *one* state in the series $\langle d|1\rangle\langle 1|s\rangle + \langle d|2\rangle\langle 2|s\rangle + \langle d|3\rangle\langle 3|s\rangle \ldots + \langle d|N\rangle\langle N|s\rangle$. A significant problem with this way of thinking is that it opens the door to the concept of a 'localized' single photon which is incompatible with the quantum principle of the non-localized photon. The non-local nature of the photon is beautifully corroborated by the

$\Delta\nu\ \Delta x \approx 1$ version of the Heisenberg's uncertainty principle and reaffirmed experimentally by linewidth measurements using Mach–Zehnder interferometers (Duarte 2003, 2014). Furthermore, this reduction is also incompatible with the utilization of the whole superposition taking into consideration every possible path (Feynman and Hibbs 1965).

As far as quantum entanglement is concerned, in the collapse scenario, the superposition probability amplitude

$$|\psi\rangle = 2^{-1/2}(|x\rangle_1\ |y\rangle_2 - |y\rangle_1\ |x\rangle_2)$$

would have to involve a reduction such as $|\psi\rangle \rightarrow |x\rangle$ in the $+z$ side of the source while reducing via $|\psi\rangle \rightarrow |y\rangle$ in the $-z$ side of the source. This collapse would have to occur in Dyson's first layer and as such it is beyond the experimentalist's measuring capability. But is this collapse necessary?

Here it should be observed that a collapse adds an additional, and unnecessary, step in the measurement process since these measurements can be described neatly in the absence of this collapse via the superposition probability approach $|\psi\rangle|\psi\rangle^* \rightarrow P(\varphi_1, \varphi_2)$.

To re-emphasize, from a broader perspective, the main argument that is being articulated in this section is that quantum measurements, in the interferometric and polarization domains, can be described without resorting to the concept of the collapse of the wave function or the collapse of the probability amplitude. These measurements can be described quite naturally via the concept of *quantum superposition probabilities* such as $\langle d|s\rangle\langle d|s\rangle^*$ and $P(\varphi_1, \varphi_2) = |\psi\rangle|\psi\rangle^*$.

10.8 Is quantum entanglement the essence of quantum mechanics?

In chapter 2 it was shown that the generalized equations for quantum entanglement applicable to n quanta and N channels of propagation ($n = N$) are (Duarte 2019, Duarte and Taylor 2017)

$$|\psi\rangle_R = N^{-1/2}\sum_{j=1}^{N}(\pm)|c\rangle_{N+1-j} \tag{10.47}$$

$$1 = ||\psi\rangle_I|^2 + ||\psi\rangle_{II}|^2 + ||\psi\rangle_{III}|^2 + ||\psi\rangle_{IV}|^2 + \ldots \tag{10.48}$$

$$|c\rangle_{N+1-j} = \prod_{m=1,3,5\ldots}^{n} |a\rangle_m\ |b\rangle_{m+1} \tag{10.49}$$

For a pair of quanta and two propagation channels ($n = N = 2$) these equations reduce to

$$|\psi\rangle = 2^{-1/2}(|x\rangle_1\ |y\rangle_2 \pm |y\rangle_1\ |x\rangle_2)$$

In Chapters 2 and 6 it was shown that the derivational path from the interferometric probability amplitude to quantum entanglement

$$\sum_{j=1}^{N} \langle d|j\rangle\langle j|s\rangle \rightarrow 2^{-1/2}(|x\rangle_1 \, |y\rangle_2 \pm |y\rangle_1 \, |x\rangle_2) \tag{10.50}$$

is natural and transparent while utilizing the Dirac identities $|c\rangle_2 = |y\rangle_1 \, |x\rangle_2$ and $|c\rangle_1 = |x\rangle_1 \, |y\rangle_2$ (Duarte 2013, 2014, 2016b).

Given the reversibility of quantum mechanical equations, the path

$$2^{-1/2}(|x\rangle_1 \, |y\rangle_2 \pm |y\rangle_1 \, |x\rangle_2) \rightarrow \sum_{j=1}^{N} \langle d|j\rangle\langle j|s\rangle \tag{10.51}$$

is also clear and transparent. However, the derivation of the quantum entanglement probability amplitude via non-interferometric physics, at this stage, appears to be somewhat removed from foundational physics principles and more related to purely mathematical procedures involving the Pauli matrices or the Hadamard matrix.

The interferometric route to quantum entanglement is direct

$$\sum_{j=1}^{N} \langle d|j\rangle\langle j|s\rangle \rightarrow 2^{-1/2}(|x\rangle_1 \, |y\rangle_2 \pm |y\rangle_1 \, |x\rangle_2)$$

while the derivational path from quantum entanglement to interference requires (see chapter 6)

$$\sum_{j} H_{ij} \, C_j \rightarrow \sigma_x, \, \sigma_y \rightarrow 2^{-1/2}(|x\rangle_1 \, |y\rangle_2 \pm |y\rangle_1 \, |x\rangle_2) \rightarrow \sum_{j=1}^{N} \langle d|j\rangle\langle j|s\rangle \tag{10.52}$$

which appears to imply that quantum interference is a more fundamental principle. This observation tends to reinforce Feynman's assertion that interference *has in it the heart of quantum mechanics* (Feynman *et al* 1965) … and Feynman wrote this sentence in his lectures even though he was perfectly aware of the physics of quantum entanglement.

Again, from an entirely neutral position, assuming that both

$$\langle d|s\rangle = \sum_{j=1}^{N} \langle d|j\rangle\langle j|s\rangle$$

and

$$|\psi\rangle = 2^{-1/2}(|x\rangle_1 \, |y\rangle_2 \pm |y\rangle_1 \, |x\rangle_2)$$

are *intrinsic foundational principles* of quantum mechanics then it is natural to conclude that these two principles are inextricably and profoundly interconnected at the most fundamental level. Although that interconnection might appear to be ineffable, mathematically it is quite transparent and occurs via Dirac's identity for indistinguishable quanta in different states $|X\rangle = |a\rangle_1 \, |b\rangle_2 \, |c\rangle_3 \ldots |g\rangle_n$ (Duarte 2013, 2016b, 2019).

10.9 On the origin of the Dirac–Feynman principle

Dirac never discussed the origin of the *bra ket* notation (Dirac 1939, 1958, Duarte 2019) and Feynman did not provide any mathematical derivation of the Dirac–Feynman principle

$$\langle d|s\rangle = \sum_{j=1}^{N}\langle d|j\rangle\langle j|s\rangle$$

All we know is that this is a powerful interferometric principle that accurately describes Nature. We also know that it works! Where does it come from?

Duarte (2013, 2014) shows that the probability amplitude of quantum entanglement $|\psi\rangle = (|x\rangle_1\,|y\rangle_2 - |y\rangle_1\,|x\rangle_2)$ can be derived from the Dirac–Feynman principle. It is also known that the equations of quantum mechanics are reversible (Duarte 2019) and that the Pauli matrices and the Hadamard matrix can be used to derive the probability amplitude of quantum entanglement independently (Duarte 2019, Duarte *et al* 2020).

Hence, as described in the previous section 10.8, i.e. equation (10.52), there is a transparent mathematical path to arrive at the Dirac–Feynman principle.

The alternative view is to go back to the extraordinary pragmatic simplicity offered by Dirac in his own description of the *bra ket* notation. From a perspective of quanta propagation it all has to do with *connectedness*. The probability amplitude to propagate from plane s to plane j is given by $\langle j|s\rangle$ and the probability amplitude to propagate from plane j to plane d is given by $\langle d|j\rangle$ (Dirac 1939, 1958) so that the superposition probability amplitude to propagate from plane s to plane d via plane j is given by $\langle d|j\rangle\langle j|s\rangle$ (Feynman *et al* 1965). That's how Nature works.

10.10 Quantum pragmatism free of paradoxes

The experiment is the ethos of physics and the measurement is the ethos of the experiment. Furthermore, as succinctly stated by van Kampen's first theorem: '*Quantum mechanics works*' (van Kampen 1988). This observation reinforces our opening sentence since quantum mechanics works as corroborated by measurements and experimental data. Since the very beginning of physics, experimental data, with measured uncertainties, has been crucial to the development of physics (Newton 1686). As far as quantum mechanics is concerned 'unless a thing can be defined by measurement, it has no place in the theory' (Feynman *et al* 1965).

In this regard, it is up to the proponents of alternative interpretations of quantum mechanics to predict new experimental phenomena so that their interpretational frameworks can be tested.

An additional point of enormous practical importance is that, besides the principles already integrated within the orthodox framework of quantum mechanics, no unique concept advanced by any of the alternative interpretations is necessary to solve quantum mechanical problems or to explain quantum mechanical measurements.

Perhaps the cleverest, most ardent, most articulate, critic of orthodox quantum mechanics was the very author of Bell's theorem: John S Bell. In this regard, it should be indicated that Bell presented no experimental evidence to back up his claims and his criticisms were erected from a philosophical framework (Bell 1966, 1990, Bell and Nauenberg 1966). Bell's criticisms against quantum mechanics can be readily dismissed (Duarte 2019).

Hidden variable theories, as suggested by Einstein and developed by Bohm, turned out to exhibit non-quantum mechanical properties (Bohm 1952a, 1952b, Bohm and Bub 1966). Furthermore, the physics of quantum entanglement as beautifully expressed in $|\psi\rangle = (|x\rangle_1 |y\rangle_2 - |y\rangle_1 |x\rangle_2)$ (Pryce and Ward 1947, Ward 1949) was developed in an absolute vacuum of philosophical arguments (Duarte 2012, 2014). To be more specific and transparent, the physics of $|\psi\rangle = (|x\rangle_1 |y\rangle_2 - |y\rangle_1 |x\rangle_2)$ is entirely independent of Bell's theorem and additional terminology in vogue within the quantum entanglement community. From a practical perspective, Bell's theorem's main impact is via the Ekert (1991) protocol used to verify that communications via $|\psi\rangle = (|x\rangle_1 |y\rangle_2 - |y\rangle_1 |x\rangle_2)$ have not been compromised by a third party or eavesdropper. In chapter 8 a simple all-quantum methodology to quantum cryptography has been introduced that is independent of Bell's theorem.

At this stage, it is unclear whether quantum interference or quantum entanglement is the most fundamental concept in quantum mechanics although there appears to be some slight edge in favor of quantum interference. What is true, however, is that these two principles are inextricably and profoundly interconnected at the most fundamental level within Dyson's first layer.

The pragmatic perspective offered in this book is entirely compatible with the views and opinions of Bohr, Born, Dirac, Dyson, Feynman, Lamb, and Ward. In particular, based on the explicit thoughts of Born, Feynman, and Lamb it is advocated that the principles of quantum mechanics, including its *indeterminacy*, ought to be embraced and applied to areas of physics thought to be well explained by classical theory. In this regard, it has already been shown that an important section of classical optics including interference → diffraction → refraction → reflection can be described, in that particular hierarchical order, starting from the Dirac–Feynman interferometric principle (Duarte 1997, 2014).

A central feature of this pragmatic perspective is that the probability amplitude for a single quantum is mathematically identical to the probability amplitude for an ensemble of indistinguishable quanta. This tends to suggest that these might be the same phenomenon.

Furthermore, from its interferometric foundations there are no paradoxes in the development of the quantum entanglement probability amplitudes. In other words, this perspective on quantum mechanics *is inherently free of paradoxes*.

Superposition probability amplitudes such as $|\psi\rangle$ are the mathematical tools of nature.

Probability amplitudes are what describe the physics. It is absolutely crucial that the physics at hand be described by the correct probability amplitudes. Quantum superposition probabilities, such as $\langle d|s\rangle\langle d|s\rangle^*$ and $|\psi\rangle|\psi\rangle^*$, accurately provide the measurables. The transition between probability amplitudes and probabilities, that is, the → in $|\psi\rangle \rightarrow |\psi\rangle|\psi\rangle^*$, is a mathematical step at the boundary. This is a transparent operation perfectly in line with the classical mathematical rules of complex numbers and $|\psi\rangle|\psi\rangle^*$ is the square of the absolute value of $|\psi\rangle$, that is $||\psi\rangle|^2$ (Flanders *et al* 1970). There is no mystery here: it should be remembered that the multiplication of complex numbers with their respective conjugates goes back at least to the times of the introduction of *Euler's formula* around 1748. Furthermore, the multiplication of complex quantities with their respective complex conjugates is an element of standard classical electromagnetic theory (Lorrain and Corson 1970, Born and Wolf 1975).

As far a *reality* is concerned: from our pragmatic quantum perspective, reality is entirely determined by measurements and experimental results… and nothing else.

Finally, a future new super quantum mechanics (SQM) might be possible but, from our perspective, this new theory will have to wait until some new unsuspected experimental phenomenon or phenomena is discovered that will provide the foundations for advancement. And yet it is highly probable that this new SQM will still be based in *indeterminacy* and *nonlocality*.

Corollary: the concept of non-locality for the photon (Lamb 1995, Dalitz and Duarte 2000) is strongly corroborated by $\Delta x \approx c/\Delta\nu$ and corresponding linewidth interferometric measurements (Duarte 2003, 2014). In the N-slit interferometric approach to the superposition probability amplitude of quantum entanglement, there is no need to struggle with or even contemplate the 'particle-wave duality.' This perspective is reinforced by Feynman's statement in reference to particle–wave ambiguities and the photon: 'We say: It is like *neither*' (Feynman *et al* 1965). From the pragmatic perspective of quantum mechanics the photon, or the quantum, is a unique quantum mechanical entity whose behavior is beautifully described by the Dirac–Feynman interferometric principle and the probability amplitude of quantum entanglement.

Besides the inherent indeterminacy, *nonlocality* of the photon is the main non-classical concept adopted in our pragmatic perspective of quantum mechanics. In other words, *locality* is discarded.

The issue of *free-will* is quite in vogue in philosophically inclined discussions. This issue does not explicitly arise in the mechanics of our pragmatic quantum perspective since our approach is independent of Bell's theorem. However, the concept of free-will is quite compatible with quantum indeterminacy and non-locality. From a macroscopic perspective, the actions of the observer, the measurer, or the experimentalist, do imply the exercise of free will via the choice of initial conditions. In his book *The Meaning of It All*, Feynman provides a valiant and eloquent defense of the *freedom to doubt* (Feynman 1998) which can only exist if there is freedom of thought.

Note: although it might be apparent to those familiar with interpretational issues, it should be stated, for the sake of completeness, that this pragmatic approach is compatible with elements of the orthodox interpretation, Born's indeterministic or statistical interpretation (Born 1949, 1954), and instrumentalism.

Problems

10.1 Work out what happens if $|P(x, y) - P(x, y')|$ becomes $|P(x, y) + P(x, y')|$.

10.2 Verify that the RHS of equation (10.9) is equal to 2.

10.3 Given the quantum probability $P(\varphi_1, \varphi_2) = -\cos 2(\varphi_1 - \varphi_2)$, calculate

$$\Sigma_P = |P(\varphi_1, \varphi_2) - P(\varphi_1, \varphi_2')| + |P(\varphi_1', \varphi_2') + P(\varphi_1', \varphi_2)|$$

for $\varphi_1 = 0$, $\varphi_2 = \pi/3$, $\varphi_1' = \pi/6$, $\varphi_2' = 0$.

10.4 Given the quantum probability $P(\varphi_1, \varphi_2) = -\cos 2(\varphi_1 - \varphi_2)$, calculate

$$\Sigma_{|\psi\rangle|\psi\rangle^*} = |P(\varphi_1, \varphi_2)| + |P(\varphi_1, \varphi_2')| + |P(\varphi_1', \varphi_2')| + |P(\varphi_1', \varphi_2)|$$

for $\varphi_1 = 0$, $\varphi_2 = \pi/3$, $\varphi_1' = \pi/6$, $\varphi_2' = 0$. Any comments? (See chapter 8).

10.5 Verify that $\langle d|s\rangle = \langle s|d\rangle^*$.

10.6 Show that $\Delta p\ \Delta x \approx h$ is equivalent to $\Delta\nu\ \Delta t \approx 1$.

10.7 Show that $\mathscr{I}(\nu) \propto \langle d|s\rangle\langle d|s\rangle^*$ can be expressed as $\mathscr{I}(\nu) = \mathrm{K}\, h\nu\langle d|s\rangle\langle d|s\rangle^*$ where K is a constant in units of $\mathrm{m}^2\ \mathrm{s}^{-1}$.

10.8 Expand equation (10.19) for $N = 2$ and provide an explicit expression for $\langle d|s\rangle\langle d|s\rangle^*$.

10.9 Verify, using numerical techniques that $\langle d|s\rangle\langle d|s\rangle^*$ yields a highly structurized interferometric distribution in the near field, lets say $D_{\langle d|j\rangle} = 10$ mm, and a smooth near-Gaussian distribution in the far field, lets say $D_{\langle d|j\rangle} = 10$ m. Use $N = 5$, $\lambda = 590$ nm, and slits 50 mm wide separated by 50 mm. Note: a program for this type of calculations is given in Duarte (2014).

10.10 Using the diffraction limited expression for beam divergence $\Delta\theta = \lambda/\pi w_b$, where $\lambda = 593.93$ nm and $2w_b = 3000$ μm, verify that the full-width of the near Gaussian beam depicted in figure 10.6 is just over 10 mm for a propagation distance of $D_{\langle d|j\rangle} = 30$ m.

10.11 Show that the generalized equation (10.47) reduces, for $n = N = 2$, to the probability amplitude for quantum entanglement given in equation (10.37).

10.12 Besides the quantum entanglement methodology described in this book and the Feynman Hamiltonian approach, also described in this book, identify at least one more avenue to independently derive the Pauli matrices from first principles.

References

Albert D Z and Vaidman L 1989 On a theory of the collapse of the wave function *Bell's Theorem, Quantum Theory and Conception of the Universe* ed M Kafatos (Berlin: Springer) pp 1–6

Andersen A, Madsen J, Reichelt C, Ahl S R, Lautrup B, Ellegaard C, Levinsen M T and Bohr T 2015 Double-slit experiment with single wave-driven particles and its relation to quantum mechanics *Phys. Rev.* E **92** 013006

Bell J S 1964 On the Einstein–Podolsky–Rosen paradox *Physics* **1** 195–200

Bell J S 1966 On the problem of hidden variables in quantum mechanics *Rev. Mod. Phys.* **38** 447–52

Bell J S 1971 Introduction to the hidden variable question *Proc. of the Int. School of Physics Enrico Fermi: Foundations of Quantum Mechanics* (New York: Academic) pp 171–81

Bell J S 1987 *Speakable and Unspeakable in Quantum Mechanics* (Cambridge: Cambridge University Press)

Bell J S 1990 Against 'measurement' *Phys. World* **3** 33–40

Bell J S and Nauenberg M 1966 The moral aspects of quantum mechanics *Preludes in Theoretical Physics* ed A De Shalit, H Feshbach and L Van Hove (Amsterdam: North Holland) pp 279–86

Bohm D 1952a A suggested interpretation of quantum theory using 'hidden' variables. I *Phys. Rev.* **85** 166–79

Bohm D 1952b A suggested interpretation of quantum theory using 'hidden' variables. II *Phys. Rev.* **85** 180–93

Bohm D and Bub J 1966 A proposed solution of the measurement problem in quantum mechanics by hidden variable theory *Rev. Mod. Phys.* **38** 453–69

Bohr N 1935 Can quantum mechanical description of physical reality be considered complete? *Phys. Rev.* **48** 696–702

Born M 1926 Zur quantenmechanik der stoßvorgänge *Z. Phys.* **37** 863–27

Born M 1949 *Natural Philosophy of Cause and Chance* (Oxford: Clarendon)

Born M 1954 The statistical interpretation of quantum mechanics *Nobel Lecture*

Born M and Wolf E 1975 *Principles of Optics* (New York: Pergamon)

Bub J 2010 von Neumann's 'no hidden variables proof': a re-appraisal *Found. Phys.* **40** 1333–40

Close F 2011 *The Infinity Puzzle* (Oxford: Oxford University Press)

Couder Y and Fort E 2006 Single-particle diffraction and interference at a microscopic scale *Phys. Rev. Lett.* **97** 154101

Dalitz R H 1987 Another side to Paul Dirac *Paul Adrien Maurice Dirac* ed B N Kursunoglu and E P Wigner (Cambridge: Cambridge University Press) ch 10

Dalitz R H and Duarte F J 2000 John Clive Ward *Phys. Today* **53** 99–100

Dirac P A M 1939 A new notation for quantum mechanics *Math. Proc. Cam. Phil. Soc.* **35** 416–8

Dirac P A M 1958 *The Principles of Quantum Mechanics* 4th edn (Oxford: Oxford University Press)

Dirac P A M 1975 Lecture delivered at the University of New South Wales, Sydney, Australia

Dirac P A M 1987 The inadequacies of quantum field theory *Paul Adrien Maurice Dirac* ed B N Kursunoglu and E P Wigner (Cambridge: Cambridge University Press) ch 15

Dombey N 2021 John Clive Ward *Biogr. Mems. Fell. R. Soc.* **70** 419–40

Duarte F J 1993 On a generalized interference equation and interferometric measurements *Opt. Commun.* **103** 8–14

Duarte F J 1997 Interference, diffraction, and refraction, via Dirac's notation *Am. J. Phys.* **65** 637–40

Duarte F J 2002 Secure interferometric communications in free space *Opt. Commun.* **205** 313–9

Duarte F J 2003 *Tunable Laser Optics* (New York: Elsevier Academic)

Duarte F J 2004 Comment on 'Reflection, refraction and multislit interference' *Eur. J. Phys.* **25** L57–8

Duarte F J 2011 *Laser Physicist* (Rochester, NY: Optics Journal)

Duarte F J 2012 The origin of quantum entanglement experiments based on polarization measurements *Eur. Phys. J.* H **37** 311–8

Duarte F J 2013 The probability amplitude for entangled polarizations: an interferometric approach *J. Mod. Opt.* **60** 1585–7

Duarte F J 2014 *Quantum Optics for Engineers* (New York: CRC Press)

Duarte F J 2016a Multiple-prism arrays and multiple-prism beam expanders *Tunable Laser Applications* ed F J Duarte (New York: CRC Press) ch 13

Duarte F J 2016b Secure space-to-space interferometric communications and its nexus to the physics of quantum entanglement *Appl. Phys. Rev.* **3** 041301

Duarte F J 2019 *Fundamentals of Quantum Entanglement* (Bristol: IOP Publishing)

Duarte F J and Taylor T S 2017 Quantum entanglement probability amplitudes in multiple channels: an interferometric approach *Optik* **139** 222–30

Duarte F J, Taylor T S, Black A M and Olivares I E 2013 Diffractive patters superimposed over propagating *N*-slit interferograms *J. Mod. Opt.* **60** 136–40

Duarte F J, Taylor T S and Slaten J C 2020 On the probability amplitude of quantum entanglement and the Pauli matrices *Opt. Quantum Electron.* **52** 106

Dyson F J 1949a The radiantion theories of Tomonaga, Schwinger, and Feynman *Phys. Rev.* **75** 486–502

Dyson F J 1949b The S matrix in quantum electrodynamics *Phys. Rev.* **75** 1736–55

Dyson F J 2007 Why is Maxwell's theory so hard to understand? *Proceedings of the 2nd European Conference on Antennas and Propagation* (Edinburgh: EuCAP)

Einstein A, Podolsky B and Rosen N 1935 Can quantum mechanical description of physical reality be considered complete? *Phys. Rev.* **47** 777–80

Ekert A K 1991 Quantum cryptography based on Bell's theorem *Phys. Rev. Lett.* **67** 661–3

Feynman R P 1965 *The Character of Physical Law* (Cambridge: MIT Press)

Feynman R P 1998 *The Meaning of it All* (Reading, MA: Addison-Wesley)

Feynman R P and Hibbs A R 1965 *Quantum Mechanics and Path Integrals* (New York: McGraw-Hill)

Feynman R P, Leighton R B and Sands M 1965 *The Feynman Lectures on Physics* vol III (Reading, MA: Addison-Wesley)

Flanders H, Korfhage R R and Price J J 1970 *Calculus* (New York: Academic)

Fraser G 2008 *Cosmic Anger* (Oxford: Oxford University Press)

Greiner W and Reinhardt J 2009 *Quantum Electrodynamics* (Berlin: Springer)

Heisenberg W 1927 Über den anschaulichen inhalt der quantentheoretishen kinematikund mechanic *Z. Phys.* **43** 172–98

Kaiser D 2005 *Drawing Theories Apart: The Dispersion of Feynman Diagrams in Postwar Physics* (Chicago, IL: University of Chicago)

Lamb W E 1986a Quantum theory of measurement *Ann. New York Acad. Sci.* **480** 407–16

Lamb W E 1986b Theory of quantum mechanical measurements *Proc. 2nd Int. Symp. on the Foundations of Quantum Mechanics (Tokyo)*

Lamb W E 1987a Sequential measurements in quantum mechanics *Quantum Measurements and Chaos* ed E R Pike and S Sarkar (New York: Plenum)

Lamb W E 1987b Schrödingers's cat *Paul Adrien Maurice Dirac* ed B N Kursunoglu and E P Wigner (Cambridge: Cambridge University Press) ch 21

Lamb W E 1995 Anti-photon *Appl. Phys.* B **60** 77–84

Lamb W E 2001 Super classical quantum mechanics: the best interpretation of nonrelativistic quantum mechanics *Am. J. Phys.* **69** 413–22

Lamb W E and Retherford R C 1947 Fine structure of the hydrogen atom by a microwave method *Phys. Rev.* **72** 241–3

Landau L D and Lifshitz E M 1977 *Quantum Mechanics* 3rd edn (New York: Pergamon)

Lorrain P and Corson D R 1970 *Electromagnetic Fields and Waves* (San Francisco, CA: Freeman)

Mandel L and Wolf E 1995 *Optical Coherence and Quantum Optics* (Cambridge: Cambridge University Press)

Namiki M and Pascazio S 1992 Many-Hilbert-spaces approach to the wave-function collapse *Found. Phys.* **22** 451–66

Newton I 1686 *Principia Mathematica* (London: Reg. Soc.)

Penrose R 1996 On gravity's role in quantum state reduction *Gen. Rel. Grav.* **28** 581–600

Planck M 1901 Ueber das gesetz der energieverteilung im normalspectrum *Ann. Phys.* **309** 553–63

Pryce M L H and Ward J C 1947 Angular correlation effects with annihilation radiation *Nature* **160** 435

Sakharov A 1990 *Memoirs* (New York: Knopf)

Schrödinger E 1944 *What Is Life?* (Cambridge: Cambridge University Press)

van Kampen N G 1988 Ten theorems about quantum mechanical measurements *Physica* A **153** 97–113

von Neumann J 1932 *Mathematische Grundlagen der Quanten-Mechanik* (Berlin: Springer)

Ward J C 1949 *Some Properties of the Elementary Particles* (Oxford: Oxford University Press)

Ward J C 1950 An identity in quantum electrodynamics *Phys. Rev.* **78** 182

Ward J C 1979 Science notes *Arena* **12** 6

Ward J C 2004 *Memoirs of a Theoretical Physicist* (Rochester, NY: Optics Journal)

IOP Publishing

F J Duarte and T S Taylor

Appendix A

More on Dirac's notation: application to laser cavities and interference

The use of Dirac's quantum notation in the description of macroscopic highly coherent high-power tunable laser sources is described. These concepts are further applied to the description of sub-microscopic electrically-pumped coherent organic semiconductor emitters. The transition from quantum to classical interference is also described.

A.1 Introduction

As mentioned in chapter 10, perhaps inspired by Born (1949), Willis E Lamb encouraged the use of quantum physics in the description of macroscopic phenomena (Lamb 1987). Indeed, Feynman applied his quantum path integrals to the description of *macroscopic diffraction* (Feynman and Hibbs 1965), a phenomenon considered by many as purely 'classical.' It was Feynman's approach to diffraction (Feynman and Hibbs 1965) and to double-slit electron interference (Feynman *et al* 1965) that inspired Duarte to extend the Dirac–Feynman principle to describe generalized N-slit interference of bosons quantum mechanically (Duarte 1993). Macroscopic experimental phenomena were also the focus of van Kampen's *Ten theorems about quantum mechanical measurements* (van Kampen 1988).

The laser is a child of the quantum era. Some lasers can be described via semi-classical approaches such as the application of classical rate equation that include quantum mechanical cross sections (Haken 1969, Sargent *et al* 1974, Duarte 2003, Silfvast 2003). Complete quantum mechanical descriptions of laser emission, or coherent emission, were first given by Lamb (1964), Feynman *et al* (1965), and Sargent *et al* (1974).

Here, quantum mechanical concepts are applied, via Dirac's notation (Dirac 1939, 1958), to the description of macroscopic intracavity laser phenomena.

doi:10.1088/978-0-7503-3407-5ch11

A.2 The cavity linewidth equation: a heuristic approach

Consider the frequency selectivity configuration of a multiple-prism grating cavity, as illustrated in figure A.1. The unfolding to the frequency selective geometry is depicted in figure A.2.

In reference to figure A.2, the double-pass propagation from s to s' can be described by (Duarte 1992)

$$\langle s'|A|s\rangle = \langle s'|\phi'_{1,m}\rangle\langle\phi'_{1,m}|A|\phi_{1,m}\rangle\langle\phi_{1,m}|s\rangle \tag{A.1}$$

and, using Born's rule (Born 1926), the corresponding probability can be written as

$$|\langle s'|A|s\rangle|^2 = |\langle s'|\phi'_{1,m}\rangle|^2|\langle\phi'_{1,m}|A|\phi_{1,m}\rangle|^2|\langle\phi_{1,m}|s\rangle|^2 \tag{A.2}$$

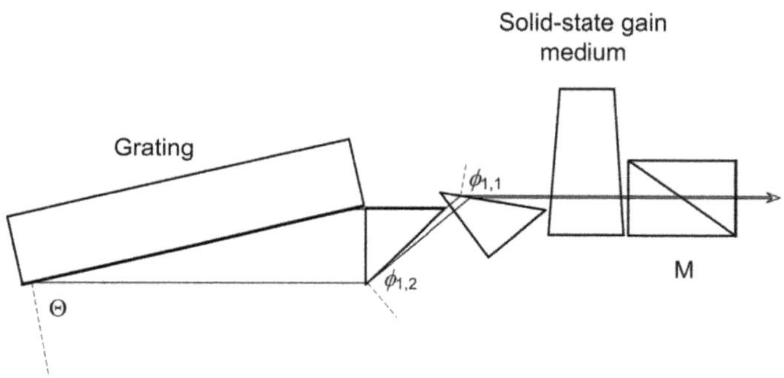

Figure A.1. Multiple-prism grating assembly in a high-performance narrow-linewidth tunable laser performing at the limit allowed by Heisenberg's uncertainty principle (from Duarte 1999). Copyright The Optical Society.

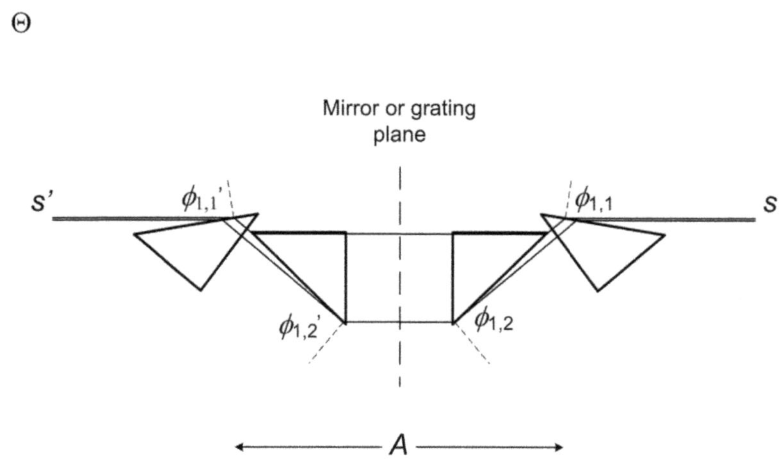

Figure A.2. Unfolded multiple-prism grating assembly. The unfolding is done about the symmetry plane provided by a mirror or diffraction grating.

In the following analysis the process considered is a mechanism of *frequency selectivity*.

That is, the physics is guided by phenomena that discriminate in favor of low beam divergence and narrow-linewidth emission. Considering that the probability of successful arrival from s to A is inversely proportional to the beam divergence $\Delta\theta$ then

$$|\langle \phi_{1,m}|s\rangle|^2 \approx \frac{\kappa_1}{\Delta\theta} \tag{A.3}$$

where κ_1 is a constant. Once the photon flux arrives at the multiple-prism grating dispersive assembly, the probability for narrow-linewidth transmission is directly proportional to the dispersion of the assembly so that

$$\left|\left\langle \phi'_{1,m}|A|\phi_{1,m}\right\rangle\right|^2 \approx \kappa_2\left(\frac{\partial\Theta}{\partial\lambda}\right) \tag{A.4}$$

For resonant, frequency selected, emission exiting A, the probability to arrive at s' is very high so that

$$\left|\left\langle s'|\phi'_{1,m}\right\rangle\right|^2 \approx 1 \tag{A.5}$$

Since the overall probability for resonant, narrow-linewidth, amplification is inversely proportional to the wavelength spread of the emission

$$|\langle s'|A|s\rangle|^2 \approx \left(\frac{\kappa_3}{\Delta\lambda}\right) \tag{A.6}$$

then, substituting equations (A.3)–(A.6) into (A.2) yields (Duarte 1992)

$$\Delta\lambda \approx \Delta\theta\left(\frac{\partial\Theta}{\partial\lambda}\right)^{-1} \tag{A.7}$$

for the special case of $\kappa_3 \approx \kappa_1\,\kappa_2$. Equation (A.7) is the well-known classical linewidth equation (Robertson 1955).

A.3 The cavity linewidth equation: an interferometric approach

As discussed in chapters 2 and 4, for single-photon propagation, or the propagation of an ensemble of indistinguishable photons, the N-slit interferometer, illustrated in figure A.3, is described directly by the Dirac–Feynman *probability amplitude*

$$\langle d|s\rangle = \sum_{j=1}^{N}\langle d|j\rangle\langle j|s\rangle \tag{A.8}$$

Again, applying Born's rule this probability amplitude leads to the one-dimensional interferometric probability

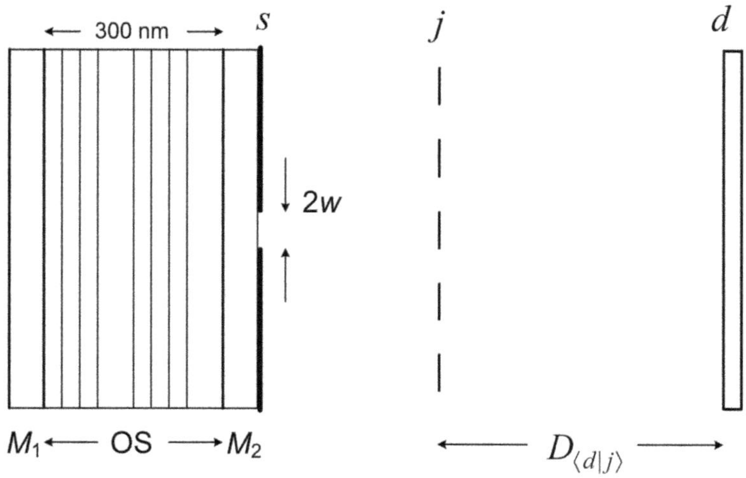

Figure A.3. *N*-slit interferometer powered by an electrically-pumped organic semiconductor coherent emitter at $\lambda \approx 540$ nm. For the experimental results discussed here $N = 2$. The 300 nm organic semiconductor gain region is labeled as OS.

$$\langle d|s\rangle\langle d|s\rangle^* = \left(\sum_{j=1}^{N}\langle d|j\rangle\langle j|s\rangle\right)\left(\sum_{j=1}^{N}\langle d|j\rangle\langle j|s\rangle\right)^* \tag{A.9}$$

that can be expressed as (Duarte 1993, 1995, 2003, 2014)

$$\langle d|s\rangle\langle d|s\rangle^* = \sum_{j=1}^{N}\Psi(r_j)\sum_{m=1}^{N}\Psi(r_m)e^{i(\Omega_m-\Omega_j)} \tag{A.10}$$

$$\langle d|s\rangle\langle d|s\rangle^* = \sum_{j=1}^{N}\Psi(r_j)^2 + 2\sum_{j=1}^{N}\Psi(r_j)\left(\sum_{m=j+1}^{N}\Psi(r_m)\cos(\Omega_m - \Omega_j)\right) \tag{A.11}$$

The phase term in this equation, $\cos(\Omega_m - \Omega_j)$, leads to (Duarte 1992, 2003, 2014)

$$d_m(\pm\sin\Theta_m \pm \sin\Phi_m) = m\lambda \tag{A.12}$$

For $\Theta_m \approx \Phi_m (= \Theta)$ this equation reduces to

$$2d\sin\Theta = m\lambda \tag{A.13}$$

which is known as the Littrow grating equation. The derivative of this equation can be written as

$$\left(\frac{\partial\Theta}{\partial\lambda}\right)\cos\Theta = \frac{m}{2d} \tag{A.14}$$

Considering two slightly different wavelengths, an expression for $(\lambda_1 - \lambda_2) = \Delta\lambda$, equation (A.13) can be written as

$$\Delta\lambda = \frac{2d}{m}(\sin\Theta_1 - \sin\Theta_2) \tag{A.15}$$

Expanding $\sin\Theta_1$ and $\sin\Theta_2$ into their corresponding series and taking the difference under the condition $\Theta_1 \approx \Theta_2(\approx\Theta)$, equation (A.15) becomes

$$\Delta\lambda \approx \frac{2d}{m}\Delta\Theta\left(1 - \frac{3\Theta^2}{3!} + \frac{5\Theta^4}{5!} - \frac{7\Theta^6}{7!}\ldots\right) \tag{A.16}$$

Next, substitution of $(m/2d)^{-1}$ into equation (A.16) yields (Duarte 1992)

$$\Delta\lambda \approx \Delta\Theta\left(\frac{\partial\Theta}{\partial\lambda}\right)^{-1}\left(1 - \frac{\Theta^2}{2!} + \frac{\Theta^4}{4!} - \frac{\Theta^6}{6!}\ldots\right)(\cos\Theta)^{-1} \tag{A.17}$$

and recognizing that the series approaches $\cos\Theta$ this equation reduces to

$$\Delta\lambda \approx \Delta\Theta\left(\frac{\partial\Theta}{\partial\lambda}\right)^{-1} \tag{A.18}$$

which has the form of the cavity linewidth equation (A.7). The difference between equations (A.7) and (A.18) is the beam divergence that is expressed as $\Delta\theta$ in equation (A.7) and as $\Delta\Theta$ in equation (A.18). The beam divergence does depend on the physics of the multiple-prism grating assembly (Duarte 2001, 2003, 2015) thus justifying the use of $\Delta\Theta$. However, in practice, the beam divergence is often measured (Duarte 2001) while the overall dispersion of the cavity is calculated. Hence, the cavity linewidth is expressed as

$$\Delta\lambda \approx \Delta\theta\left(\frac{\partial\Theta}{\partial\lambda}\right)^{-1}$$

or

$$\Delta\lambda \approx \Delta\theta(\nabla_\lambda\Theta)^{-1} \tag{A.19}$$

where $\nabla_\lambda = \partial/\partial\lambda$.

This cavity linewidth equation is widely applied to estimate the laser linewidth in pulsed high- power narrow-linewidth tunable lasers (Duarte 1990, 1999, Duarte and Piper 1984, Duarte and Conrad 1987). These narrow-linewidth tunable lasers have found wide applicability in a plethora of fields including laser guide star for astronomy (Everett 1989, Primmerman *et al* 1991, Bass *et al* 1992, Duarte 2001), laser isotope separation (Duarte and Piper 1984, Murray 1987, Bass *et al* 1992, Duarte 2016), and laser spectroscopy (Demtröder 2003).

A.4 Coherent electrically-pumped organic interferometric emitters

Coherent emission from electrically-pumped organic semiconductors was reported by Duarte *et al* (2005) while utilizing a coumarin 545 T laser dye-doped tandem organic light emitting diode (OLED) excited in the pulsed regime.

This coherence is characterized by highly directional emission, with a near diffraction-limited beam at $\Delta\theta \approx 2.53$ mrad, and high visibility interferometric fringes with $\mathscr{V} \approx 0.9$ (Duarte *et al* 2005, Duarte 2007, 2008). Further details are given in Duarte and Vaeth (2018).

Coherent emission from the electrically-excited laser dye-doped organic semiconductor was characterized in mainly two ways: directionality of the emission and the high visibility of the interference fringes of the emission. The gain medium is the laser dye coumarin 545 tetramethyl (Duarte *et al* 2006). A schematic of the experimental configuration is depicted in figure A.3.

A highly directional emission beam was recorded on black and white silver-halide film as depicted in figure A.4. The near-Gaussian digital profile of this emission is shown in figure A.5. This directional emission was measured to exhibit a beam divergence of $\Delta\theta \approx 2.53 \pm 0.13$ mrad which translates into $\Delta\theta \approx 1.09(\lambda/\pi w)$, where $(\lambda/\pi w)$ is the diffraction limited beam divergence and $\lambda \approx 540$ nm. This class of single-transverse-mode near-Gaussian beam profile, and measured beam divergence, is typical of optimized narrow-linewidth laser oscillator emission (Duarte 1999).

Figure A.6 shows the digital image of the profile of the double-slit interferogram of the highly directional on-axis emission from the electrically-pumped laser-dye doped organic semiconductor emitter. The width of the slits is 50 μm and the slits are also separated by 50 μm. The measured visibility (Michelson 1927)

$$\mathscr{V} = \frac{I_{\max} - I_{\min}}{I_{\max} + I_{\min}} \tag{A.20}$$

of this emission is $\mathscr{V} \approx 0.901 \pm 0.088$. The black and white silver-halide photographic version of figure A.6 is shown in figure A.7.

Figure A.8 shows the digital image of the profile of the double-slit interferogram of the highly directional on-axis emission from the $(3s_2 - 2p_{10})$ transition of a He–Ne laser, at $\lambda = 543.30$ nm. This coherent emission is registered under an identical interferometric configuration as that utilized to record the results displayed in figure A.6. The measured visibility of this emission is $\mathscr{V} \approx 0.952 \pm 0.031$ (Duarte 2007).

A.4.1 Coherence

The emission measurements presented in the previous section are characterized by a single-transverse-mode nearly diffraction limited beam divergence $\Delta\theta \approx 1.09\ (\lambda/\pi w)$, with near-Gaussian beam profile, and an interferometric visibility of $\mathscr{V} \approx 0.901 \pm 0.088$. These characteristics point unequivocally toward strong spatial coherence and strong spectral coherence.

Is there an explanation for the spatial coherence from a classical perspective? Collett–Wolf sources are partially-coherent sources known to produce Gaussian-like beams that are also directional (Collett and Wolf 1978). However, there is *no record* in the published literature of Collett–Wolf sources yielding interferometric visibilities in the laser range, that is, in the $0.85 \leqslant \mathscr{V} \leqslant 1.0$ range.

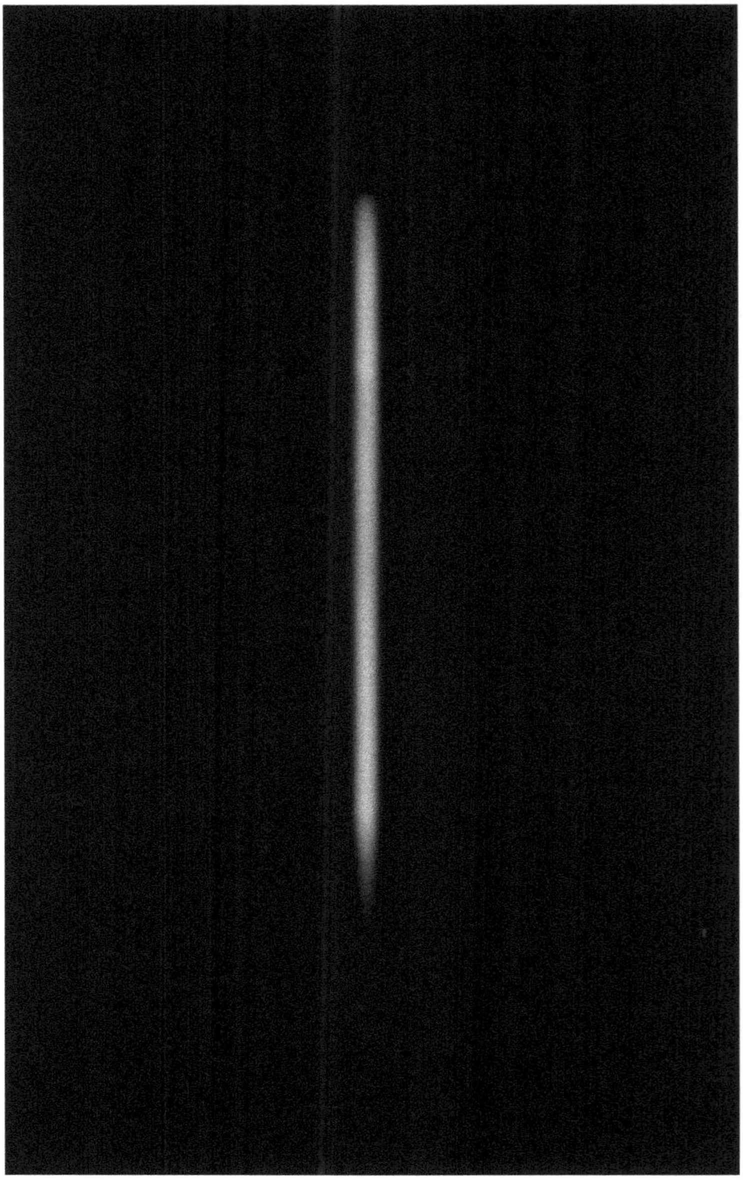

Figure A.4. Silver halide photographic image of the cross section of the on-axis emission beam from the electrically-pumped laser-dye doped organic semiconductor (from Duarte *et al* 2005). Copyright The Optical Society.

It should be noted that in the literature, emission associated with interferometric visibilities in the range $0.85 \leqslant \mathscr{V} \leqslant 1.0$ is considered as *laser emission* (Shimkaveg *et al* 1992, Trebes *et al* 1992, Ditmire *et al* 1996, Lucianetti *et al* 2004). In contrast, researchers working on semi coherent and non-laser sources including standard Alql3 OLED devices, and amplified spontaneous emission sources, report interferometric

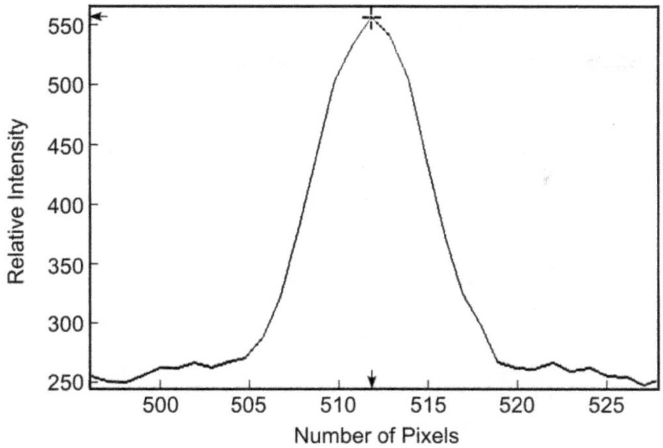

Figure A.5. Digital profile of the cross section of the on-axis emission beam from the electrically-pumped laser-dye doped organic semiconductor corresponding to figure A.5. On the horizontal axis each pixel is 25 μm wide (from Duarte *et al* 2005). Copyright The Optical Society.

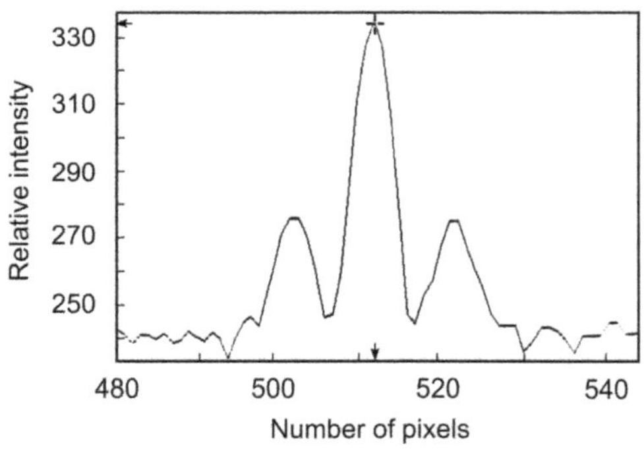

Figure A.6. Digital image of the double-slit interferogram of the highly directional on-axis emission from the electrically-pumped laser-dye doped organic semiconductor emitter. On the horizontal axis each pixel is 25 μm wide (from Duarte *et al* 2005). Copyright The Optical Society.

visibilities in the range of $0.40 \leqslant \mathscr{V} \leqslant 0.65$ (Thompson and Wolf 1957, Saxena *et al* 2006, Dharmadhikari *et al* 2005).

Since the gain medium length is $L \approx 300$ nm, and emission wavelength $\lambda \approx 540$ nm, the cavity is a sub-microcavity operating in the $L \leqslant \lambda$ regime where laser thresholds decrease dramatically. Indeed, De Martini and Jakobovitz (1988) reports on a near 'zero-threshold-laser' for a cavity length of $L \approx \lambda/2$. This observation was made by these authors while experimenting on a multiple-transverse-mode optically-pumped dye laser.

Assuming perfect coherent illumination either from a single photon or *an ensemble of indistinguishable photons* the interferometric spatial distribution can be calculated from the probabilities based on the Dirac–Feynman principle (Dirac 1958, Feynman *et al* 1965), that is, equations (A.9)–(A.11) (Duarte 1993, 1995). The interferogram thus calculated is displayed in figure A.9. In this theoretical

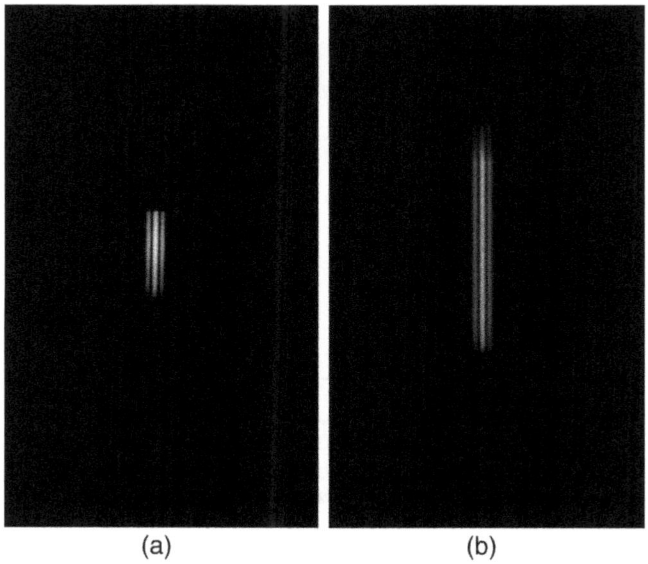

Figure A.7. Silver halide photographic image of the double-slit interferogram of the highly directional on-axis emission from (a) high-power emission from a coumarin 545 T laser and (b) the electrically-pumped laser-dye doped organic semiconductor emitter (from Duarte *et al* 2005). Copyright The Optical Society.

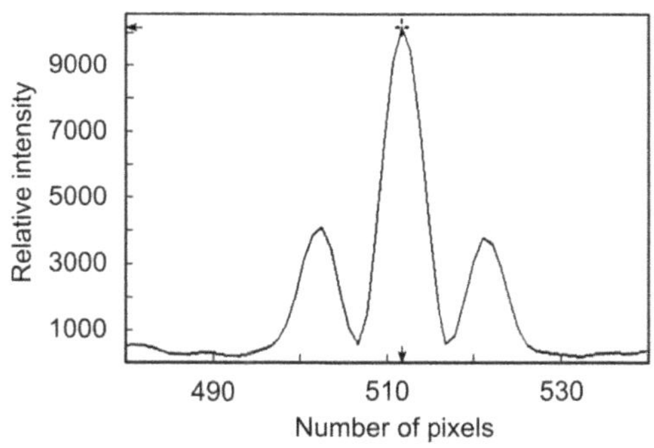

Figure A.8. Digital image of the double-slit interferogram of the highly directional on-axis emission from the $(3s_2 - 2p_{10})$ transition of a He–Ne laser emission under an identical interferometric configuration. On the x-axis each pixel is 25 μm wide (from Duarte *et al* 2005). Copyright The Optical Society.

interferogram the separation of the two secondary peaks is $\Delta d \approx 490$ μm which should be compared to $\Delta d \approx 484$ μm from the measured interferogram depicted in figure A.6. The separation of the two secondary peaks for the $(3s_2 - 2p_{10})$ transition from the He–Ne laser is measured at $\Delta d \approx 485$ μm.

A.4.2 The quantum perspective

As highlighted in previous publications (Duarte 2013, 2014, 2019), crucial to the derivation of the quantum entanglement probability amplitude (Pryce and Ward 1947, Ward 1949)

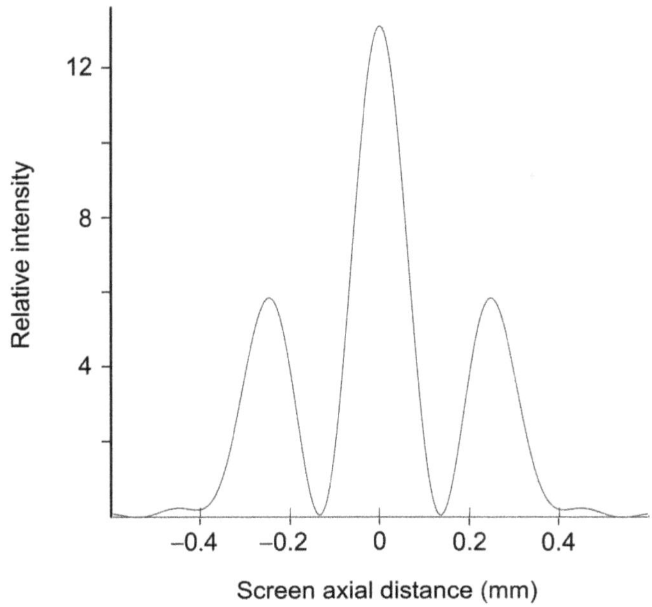

Figure A.9. Calculated profile of the double-slit interferogram. Here, $N = 2$, slit width and slit separation is 50 μm, $\lambda = 540$ nm, and the intra interferometric distance is $D_{\langle d|j\rangle} = 50$ mm.

$$|\psi\rangle = 2^{-1/2}(|x\rangle_1\, |y\rangle_2 \pm |y\rangle_1\, |x\rangle_2) \tag{A.21}$$

via the Dirac–Feynman interferometric principle (Dirac 1958, Feynman *et al* 1965), is the Dirac identity (Dirac 1958)

$$|X\rangle = |a\rangle_1\, |b\rangle_2\, |c\rangle_3 \ldots |g\rangle_n \tag{A.22}$$

which means that indistinguishable quanta 1, 2, 3, … , n can be in different states such as $|a\rangle$, $|b\rangle$, $|c\rangle$…. When referring to this 'curious phenomena … having no analogue in classical theory' Dirac is writing about quanta of the '*same kind*' and '*absolutely indistinguishable from one another*' (Dirac 1958). At the same time he is also contemplating photon 1 in different states such as $|a\rangle_1$, $|b\rangle_1$, $|c\rangle_1$…, photon 2 in $|a\rangle_2$, $|b\rangle_2$, $|c\rangle_2$…, photon 3 in $|a\rangle_3$, $|b\rangle_3$, $|c\rangle_3$…, and so on. Thus, in analogy to $|X\rangle = |a\rangle_1, |b\rangle_2, |c\rangle_3 \ldots |g\rangle_n$ the following identities can be written

$$|Y\rangle = |a\rangle_1\, |b\rangle_1\, |c\rangle_1 \ldots |g\rangle_1 \tag{A.23}$$

$$|Z\rangle = |a\rangle_2\, |b\rangle_2\, |c\rangle_2 \ldots |g\rangle_2 \tag{A.24}$$

$$|\xi\rangle = |a\rangle_3\, |b\rangle_3\, |c\rangle_3 \ldots |g\rangle_3 \tag{A.25}$$

and so on. For bosons, Dirac tabulates the various possible series of states in the determinant

$$\begin{vmatrix} |a\rangle_1 & |a\rangle_2 & |a\rangle_3 & \cdot & \cdot & \cdot & |a\rangle_n \\ |b\rangle_1 & |b\rangle_2 & |b\rangle_3 & \cdot & \cdot & \cdot & |b\rangle_n \\ |c\rangle_1 & |c\rangle_3 & |c\rangle_3 & \cdot & \cdot & \cdot & |c\rangle_n \\ \vdots & \vdots & \vdots & \vdots & \vdots & \vdots & \vdots \\ |g\rangle_1 & |g\rangle_2 & |g\rangle_3 & \cdot & \cdot & \cdot & |g\rangle_n \end{vmatrix} \tag{A.26}$$

In Dirac's discussion on bosons there is no barrier to describing single-transverse-mode single-longitudinal-mode coherent emission of indistinguishable quanta including both, $|x\rangle$ and $|y\rangle$, states of polarization. In this regard, for $|x\rangle$ polarization, one can have

$$|\alpha\rangle = |x\rangle_1\, |x\rangle_2\, |x\rangle_3 \ldots |x\rangle_n \tag{A.27}$$

and for $|y\rangle$ polarization

$$|\beta\rangle = |y\rangle_1\, |y\rangle_2\, |y\rangle_3 \ldots |y\rangle_n \tag{A.28}$$

These identities describe precisely and elegantly what coherent emission is at a fundamental level: ensembles of indistinguishable quanta all in a given state of polarization.

Ensemble states of indistinguishable quanta, such as $|\alpha\rangle$ and $|\beta\rangle$, are precisely the class of state that can give origin to the high degree of coherence observed in the double-slit interferogram depicted in figure A.6. This perspective is in line with Dirac's assertion that '*with Bose statistics the probability of two particles being in the same state is greater than with classical statistics*' (Dirac 1958). The statistics of states with n insistinguishable bosons is discussed by Feynman *et al* (1965).

A.5 Quantum interference and classical interference

Here, the difference between the mathematical formalisms that apply to classical interference (Steel 1967) and to quantum interference (Feynman *et al* 1965, Duarte 1993) are clearly stipulated. This exposition follows the review given in Duarte (2019).

A.5.1 Classical interference

Using Maxwell's electromagnetic theory the interference equation for two-slit interference, or Young's interference, can be derived as (Born and Wolf 1999)

$$I = I_1 + I_2 + 2(I_1 I_2)^{1/2} \cos\delta \tag{A.29}$$

where I_1 and I_2 are *intensities*. An alternative notation for double-slit interference is provided by Michelson (1927)

$$i = a_1^2 + a_2^2 + 2a_1 a_2 \cos\delta \tag{A.30}$$

where a_1 and a_2 are designated as amplitudes. For an array of N-slits, the interference equation (A.29) can be extended to (Duarte 2019)

$$I = \sum_{n=1}^{N} I_n + 2\sum_{n=1}^{N} I_n^{\frac{1}{2}}\left(\sum_{m=n+1}^{N} I_m^{\frac{1}{2}}\cos\delta\right) \tag{A.31}$$

where, I_1, I_2, I_3, … , I_n are the intensities associated with each of the 1, 2, 3…, n slits. In this equation δ is the phase angle derived from the interaction of the light illuminating the slits and the geometry of the slits.

Equations (A.29)–(A.31) are classical intensity equations, these are not probability equations, and these equations are not applicable to describe single-photon interference.

A.5.2 Quantum interference

As discussed in chapters 2 the N-slit interferometer is perfectly described by the explicit version of *generalized interferometric probability* in one dimension

$$|\langle d|s\rangle|^2 = \sum_{j=1}^{N}\Psi(r_j)^2 + 2\sum_{j=1}^{N}\Psi(r_j)\left(\sum_{m=j+1}^{N}\Psi(r_m)\cos(\Omega_m - \Omega_j)\right) \tag{A.32}$$

This interferometric probability applies to single-photon propagation or to the propagation of *ensembles of indistinguishable photons*. This equation can also be expressed in two and three dimensions as given in chapter 2 (Duarte 2019).

Equations (A.9)–(A.11) and (A.32) are quantum probability equations, these are not intensity equations. These equations describe single-photon interference and interference of populations of indistinguishable photons.

For these equations to apply to semi-coherent, or broadband emission, they have to be adapted to the conditions of semi-coherent physics, as described in chapter 2.

As discussed in chapter 2, for an ensemble of r indistinguishable quanta the quantum probabilities are related to the measurable intensities by equations of the form

$$\mathscr{I}(\nu)_r = r\mathrm{K}\, h\nu|\langle d|s\rangle\langle d|s\rangle^*| \tag{A.33}$$

In this equation it is the dimensionless $|\langle d|s\rangle\langle d|s\rangle^*|$ that provides all the spatial information. Also, the intensity $\mathscr{I}(\nu)$ has the units of J s^{-1} m^{-2} or W m^{-2}.

Following Duarte (2019) the difference between classical and quantum interference can be stated quite succinctly.

Classical interference:

$$\text{Maxwell equations} \rightarrow \sum_{n=1}^{N} I_n + 2\sum_{n=1}^{N} I_n^{\frac{1}{2}}\left(\sum_{m=n+1}^{N} I_m^{\frac{1}{2}}\cos\delta\right) \tag{A.34}$$

Quantum interference:

$$\sum_{j=1}^{N}\langle x|j\rangle\langle j|s\rangle \rightarrow \sum_{j=1}^{N}\Psi(r_j)\sum_{m=1}^{N}\Psi(r_m)e^{i(\Omega_m - \Omega_j)} \tag{A.35}$$

A.5.3 From quantum to classical

The single-photon probability

$$\langle d|s\rangle\langle d|s\rangle^* = \left(\sum_{j=1}^{N}\langle d|j\rangle\langle j|s\rangle\right)\left(\sum_{j=1}^{N}\langle d|j\rangle\langle j|s\rangle\right)^*$$

and its corresponding intensity

$$\mathscr{I}(\nu) = \mathrm{K}\, h\nu|\langle d|s\rangle\langle d|s\rangle^*|$$

are *purely quantum*. The intensity for an ensemble of r indistinguishable quanta

$$\mathscr{I}(\nu)_r = r\mathrm{K}\, h\nu|\langle d|s\rangle\langle d|s\rangle^*|$$

is also a quantum quantity. For instance, $\mathscr{I}(\nu)_r$ is the intensity of a highly-coherent, or a highly-monochromatic, beam of light passing through a single slit that can be treated mathematically as being comprised of hundreds of imaginary sub slits (Duarte 1993).

In the classical domain, where intensities are measured, a transition occurs, or an equivalence is permitted,

$$\mathscr{I}(\nu)_r \to I_n \tag{A.36}$$

thus allowing classical interference of coherent intensities via

$$I = \sum_{n=1}^{N} I_n + 2\sum_{n=1}^{N} I_n^{\frac{1}{2}}\left(\sum_{m=n+1}^{N} I_m^{\frac{1}{2}}\cos\delta\right)$$

A.6 Intracavity double-mode beating instability

High-power narrow-linewidth tunable lasers can yield emission confined within a single-transverse-mode *and* a single-longitudinal-mode (Duarte and Conrad 1986, Duarte 1999).

Under these circumstances the temporal emission is characterized by a smooth near-Gaussian pulse, $\Delta t \approx 3$ ns at full-width half-maximum (FWHM), as illustrated in figure A.10. In the frequency domain the laser linewidth corresponding to that near Gaussian pulse is $\Delta\nu \approx 350$ MHz. The interferogram of this emission is shown in figure A.11.

High-laser-emission performance of this class is near the limit imposed by Heisenberg's uncertainty principle since $\Delta\nu\Delta t \approx 1.05$.

When two longitudinal modes, rather than a single-longitudinal mode, are present in the laser emission, the two modes beat against each other, as illustrated in figure A.12. In other words, they interfere according to the classical interferometric equation (A.34) which for $N = 2$ can be expressed as

$$I(\nu, t) = I_1(\nu, t) + I_2(\nu, t) + 2I_1(\nu, t)^{1/2}I_2(\nu, t)^{1/2}\cos 2\pi(\nu_2 - \nu_1)t \tag{A.37}$$

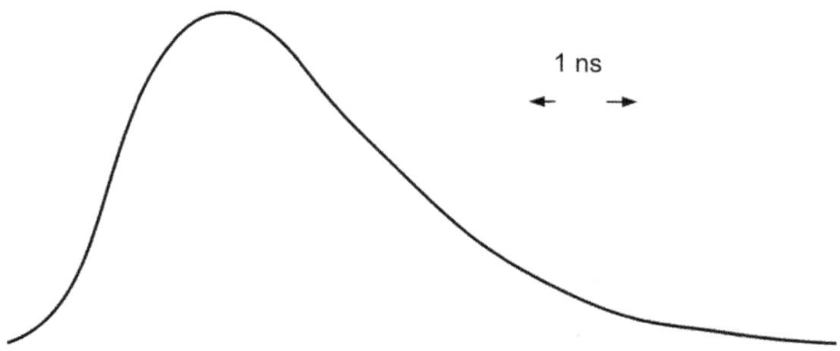

Figure A.10. Near-Gaussian temporal profile corresponding to single-longitudinal-mode emission from the dispersive laser oscillator depicted in figure A.1. The temporal scale is 1 ns/div., from Duarte (1999). Copyright The Optical Society.

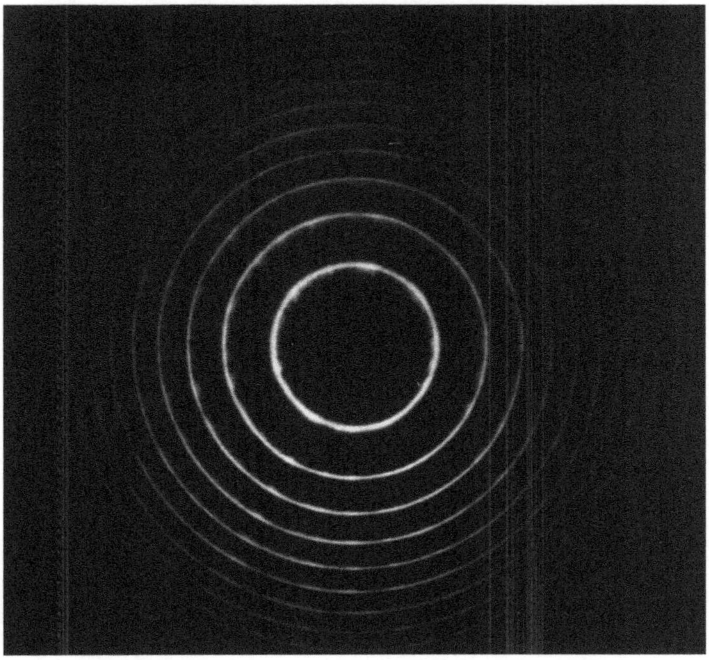

Figure A.11. Fabry–Perot interferogram corresponding to single-longitudinal-mode emission, from the dispersive laser oscillator depicted in figure A.1, at a linewidth of $\Delta\nu \approx 350$ MHz, from Duarte (1999). Copyright The Optical Society.

For two uniform and stable intensities $I_1(\nu, t)$ and $I_2(\nu, t)$, equation (A.37) yields a beautiful stable and *unmodulated* double-mode oscillation, as shown in figure A.13, which reproduces remarkably well the measured stable double-mode of oscillation shown in figure 3 of Duarte *et al* (1988).

However, when the relative intensity of the modes varies during the duration of the emission this uniform oscillation becomes unstable, or *modulated*, as illustrated in figure A.12. The duration of the laser pulse shown in this figure is $\Delta t \approx 100$ ns

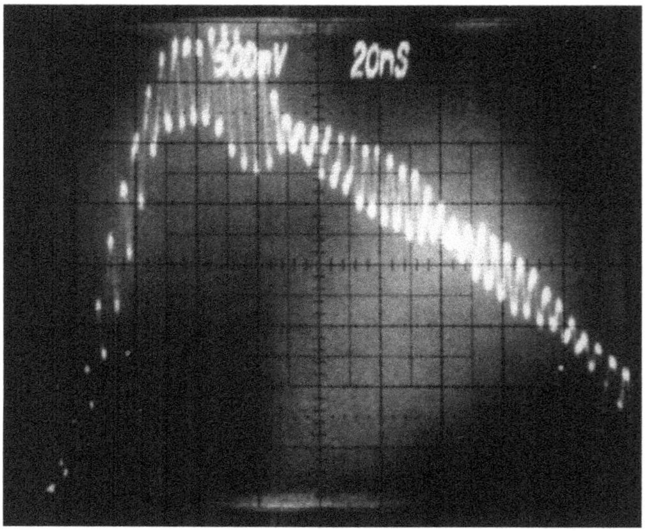

Figure A.12. Measured double-mode beating oscillation corresponding to double-longitudinal-mode oscillation, at a mode spacing of $\delta\nu \approx 250$ MHz, with a superimposed lower frequency modulation. The temporal scale is 20 ns/div., from Duarte *et al* (1988). Copyright The Optical Society.

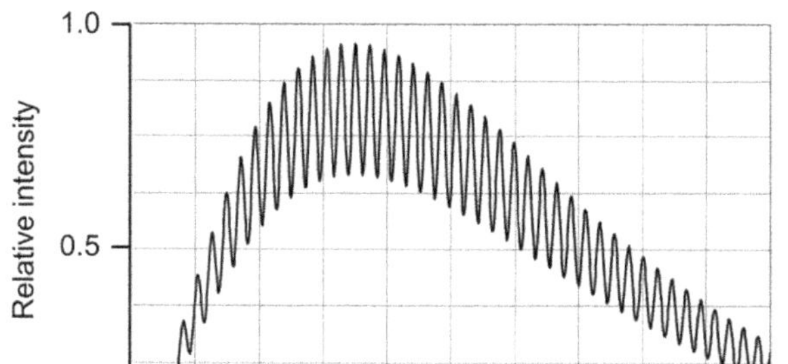

Figure A.13. Calculated stable double-mode beating oscillation corresponding to $\delta\nu = (\nu_2 - \nu_1) \approx 250$ MHz.

(FWHM) and the intracavity mode separation is $\delta\nu = (\nu_2 - \nu_1) \approx 250$ MHz (Duarte *et al* 1988).

This interferometric temporal measurement can be approximated using for the amplitudes a non-Gaussian temporal representation, derived from experimental data, of the form

$$f(t) = (a_2 t^2 + a_1 t + a_0)(b_1 t + b_0)^{-1} \tag{A.38}$$

and a ratio of 'frequency jitter' $\delta\nu_j$, to intracavity mode spacing $\delta\nu$, represented by a sinusoidal function at a frequency of 20 MHz. The resulting modulated double-mode

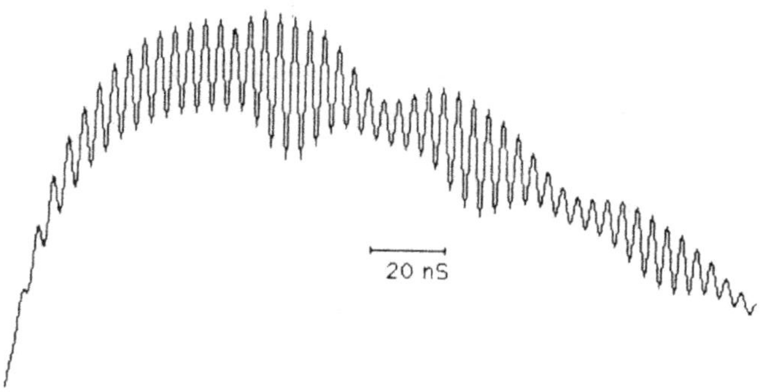

Figure A.14. Calculated double-mode beating oscillation assuming interference between the two intracavity modes and a frequency jitter of these modes, see text, that are assumed to give origin to the lower frequency modulation of the interference, from Duarte *et al* (1988). Copyright The Optical Society.

oscillation can then be calculated as illustrated in figure A.14. The initial intensity ratio for the mode at ν_1 to the mode at ν_2 is 200:1.

The measured interferogram is indicative of an emblematic double-mode oscillation with a superimposed lower frequency modulation. As suggested, the lower frequency modulation of the mode beating can be explained as a result of frequency jitter. The physics behind this perspective is a fixed frequency gain window with a near-Gaussian profile, established by the intracavity dispersion of the multiple-prism grating assembly (Duarte 2003), while the frequency position of the two intracavity modes oscillate in tandem relative to the central frequency so that their respective intensities vary as a function of time.

It should be noted that usually interferometric equations of the form of equation (A.37) are expressed in terms of the electric field E_1 and E_2 (Crawford 1968, Pacala *et al* 1984, Duarte *et al* 1988).

A.6.1 Alternative double-mode instability characterization

Comparison of figures A.12 and A.13 indicates that the physical model based on the classical double-mode interference approach, equation (A.37), is insufficient to describe the type of dual-longitudinal mode oscillation observed in the narrow-linewidth flashlamp pumped dye laser cavity (Duarte *et al* 1988). One probable explanation is that of frequency jitter as just given above. Here we explore an alternative explanation.

From the measurement shown in figure A.12 it is clear that the two intracavity modes exhibit a lower frequency oscillation superimposed over them. This unstable double-mode oscillation can be approximated alternatively by considering two extra distinguishable intracavity *rogue modes*. These rogue modes might originate from amplified spontaneous emission (ASE) which is a well-known phenomenon in narrow-linewidth high-power and high-energy tunable lasers (Duarte 1983, Duarte and Conrad 1987).

Abstracting the (ν, t) dependence and adding the two extra modes, the intensity equation can now be expressed as

$$I = I_1 + I_2 + 2I_1^{1/2}I_2^{1/2}\cos 2\pi(\nu_2 - \nu_1)t + I_3\cos^2 2\pi\upsilon_3 t + I_4\cos^2 2\pi\nu_4 t \tag{A.39}$$

The modified oscillation generated by this equation is depicted in figure A.15 that should be compared with the measurement shown in figure A.12. It is not until two new oscillating signals are added that the larger oscillating envelope or instability is seen in the physical model. This suggests that there are interactions due to the photons at these specific frequencies that might be acting parasitically within the cavity. These two added low-intensity rogue modes are distinguishable from the original two lasing modes and appear to be usurping gain from them. An additional required adjustment to the model is that with the two new added rogue modes the overall intracavity mode separation within the model must be adjusted in order to better approximate the experimental data. Note that the original model has a mode separation of $\delta\nu \approx 250$ MHz while for this alternative model $\delta\nu \approx 120$ MHz.

In summary: the experimental reality exhibited in the measurement depicted in figure A.12 can be approximated by two probable phenomena, that of frequency jitter and that based on the appearance of rogue low-intensity modes. Even in the classical domain explaining physical reality, i.e. a measurement, can be limited by uncertainty.

A.6.2 Mode intensities

The relative intensities in these calculations can be replaced by measurement compatible units of J s^{-1} m^{-2} or W m^{-2} using a representation of the intensities (see chapter 2) so that

$$I_1(\nu, t) = \alpha_1 n_1 h\nu_1 f_1(t)^{-1} \tag{A.40}$$

$$I_2(\nu, t) = \alpha_2 n_2 h\nu_2 f_2(t)^{-1} \tag{A.41}$$

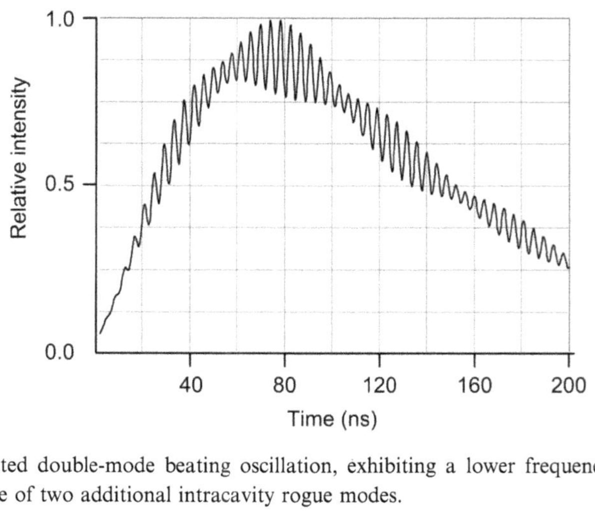

Figure A.15. Calculated double-mode beating oscillation, exhibiting a lower frequency modulation, while assuming the presence of two additional intracavity rogue modes.

where $\alpha_1 = (4\pi^2\sigma_1)^{-1}$ and $\alpha_2 = (4\pi^2\sigma_2)^{-1}$ are emission constants in units of m^{-2}, n_1 and n_2 are the number of photons in each mode, ν_1 and ν_2 are the corresponding frequencies of oscillation in Hz, h is Planck's constant, and $f_1(t)^{-1}$ and $f_2(t)^{-1}$ are temporal functions derived from grey body radiation.

A.7 Dirac's 'coincidences'

In this final section we return to a topic considered in chapter 1: Planck's constant. In his discussion on cosmology Dirac introduced units of time whose ratios lead to (Dirac 1938)

$$\alpha^{-1} = h\left(\frac{2c\varepsilon_0}{e^2}\right) \approx 137 \tag{A.42}$$

In the same paper Dirac wrote about 'coincidences' and that numbers occurring in Nature are '*connected*'.

Dirac connected quantum and Newtonian forces to arrive at his large number hypothesis (LNH) (Dirac 1938). Following Dirac's lead and connecting quantum and Newtonian energies, while considering a single photon at the edge of the *known* universe, we can arrive at

$$h \approx \alpha^{-1}\left(\frac{e^2}{2\varepsilon_0}\right)\left(\frac{r_U}{M_U G}\right)^{1/2} \tag{A.43}$$

thus providing a possible outline of the profound significance of Planck's constant h.

Equation (A.43) can be rearranged to provide

$$r_U \approx GM_U\left(\hbar\alpha\left(\frac{e^2}{4\pi\varepsilon_0}\right)^{-1}\right)^2 \tag{A.44}$$

Using $G = 6.67384 \times 10^{-11}$ m^3 kg^{-1} s^{-2}, $M_U = 1.5 \times 10^{53}$ kg, $h = 6.62607015 \times 10^{-34}$ Js, $\alpha^{-1} = 137.035999710$, and $(e^2/4\pi\varepsilon_0) = 2.3070774 \times 10^{28}$ N m^2 yields a radius for the observable universe of $r_U \approx 1.1 \times 10^{26}$ m which should be compared with ~8.5×10^{26} m as obtained from formal general relativity methodology (Lemaître 1931). It seems an astounding coincidence indeed that a rudimentary exercise *a là Dirac* yields a cosmological number with the correct order of magnitude. Some of today's cosmologists tend to dismiss Dirac's approach as 'numerology', however, experimental physics *is* about numbers.

Problems

A.1 Show that equation (A.12) reduces to the Littrow grating equation (A.13) for $\Theta_m \approx \Phi_m(= \Theta)$.

A.2 Show that using the geometrical identity

$$(\sin\Theta_1 - \sin\Theta_2) = 2\sin\frac{(\Theta_1 - \Theta_2)}{2}\cos\frac{(\Theta_1 + \Theta_2)}{2}$$

under the approximation $\Theta_m \approx \Phi_m(=\Theta)$, in equation (A.15), also leads to the cavity linewidth equation (A.18).

A.3 Expand equation (A.31) for $N = 2$ and show that it is equation (A.29).

A.4 Setting $I_1 = E_1^2$ and $I_2 = E_2^2$ in equation (A.31) show that it can be expressed as equation (A.37).

References

Bass I L, Bonnano R E, Hackel R P and Hammond P R 1992 High-average-power dye laser at Lawrence Livermore National Laboratory *Appl. Opt.* **31** 6993–7006

Born M 1926 Zur quantenmechanik der stoßvorgänge *Z. Phys.* **37** 863–827

Born M 1949 *Natural Philosophy of Cause and Chance* (Oxford: Oxford University)

Born M and Wolf E 1999 *Principles of Optics* 7th edn (Cambridge: Cambridge University)

Collett E and Wolf E 1978 Is complete spatial coherence necessary for the generation of highly directional light beams? *Opt. Lett.* **2** 27–9

Crawford F S 1968 *Waves* (New York: McGraw-Hill)

De Martini F and Jakobovitz J R 1988 Anomalous spontaneous-emission-decay phase transition and zero-threshold laser action in a microscopic cavity *Phys. Rev. Lett.* **60** 1711–4

Demtröder W 2003 *Laser Spectroscopy* 3rd edn (Berlin: Springer)

Dharmadhikari J A, Dharmadhikari A K and Kumar G R 2005 High-contrast interference pattern of amplified spontaneous emission from dyes under transient grating excitation *Opt. Lett.* **30** 765–7

Dirac P A M 1938 A new basis for cosmology *Proc. R. Soc.* **A165** 199–208

Dirac P A M 1939 A new notation for quantum mechanics *Math. Proc. Cambridge Phil. Soc.* **35** 416–8

Dirac P A M 1958 *The Principles of Quantum Mechanics* 4th edn (Oxford: Oxford University)

Ditmire T, Gumbrell E T, Smith R A, Tisch J W G, Meyerhofer D D and Hutchison M H R 1996 Spatial coherence measurements of soft x-ray radiation produced by high-order harmonic generation *Phys. Rev. Lett.* **77** 4756–9

Duarte F J 1983 Thermal effects in double prism dye laser cavities *IEEE J. Quantum Electron.* **QE-19** 1345–7

Duarte F J 1990 Pulsed dye laser apparatus for high PRF operation *US Patent* 4891817

Duarte F J 1992 Cavity dispersion equation $\Delta\lambda \approx \Delta\theta(\partial\theta/\partial\lambda)^{-1}$: a note on its origin *Appl. Opt.* **33** 6978–82

Duarte F J 1993 On a generalized interference equation and interferometric measurements *Opt. Comm.* **103** 8–14

Duarte F J 1995 Interferometric imaging *Tunable Laser Applications* ed F J Duarte (New York: Marcel Dekker) ch 5

Duarte F J 1999 Multiple-prism grating solid-state dye laser oscillator: optimized architecture *Appl. Opt.* **38** 6347–9

Duarte F J 2001 Multiple-return-pass beam divergence and the linewidth equation *Appl. Opt.* **40** 3038–41

Duarte F J 2003 *Tunable Laser Optics* (New York: Elsevier Academic)

Duarte F J 2007 Coherent electrically excited organic semiconductors: visibility of interferograms and emission linewidth *Opt. Lett.* **32** 412–4

Duarte F J 2008 Coherent electrically-excited organic semiconductors: coherent or laser emission? *Appl. Phys.* B **90** 101–8

Duarte F J 2013 The probability amplitude for entangled polarizations: an interferometric approach *J. Mod. Opt.* **60** 1585–7
Duarte F J 2014 *Quantum Optics for Engineers* (New York: CRC)
Duarte F J 2015 *Tunable Laser Optics* 2nd edn (New York: CRC)
Duarte F J 2016 *Tunable Laser Applications* (New York: CRC)
Duarte F J 2019 *Fundamental of Quantum Entanglement* (Bristol: IOP Publishing)
Duarte F J and Conrad R W 1986 Single-mode flashlamp-pumped dye laser oscillators *Appl. Opt.* **25** 663–5
Duarte F J and Conrad R W 1987 Diffraction-limited single-longitudinal-mode multiple-prism flashlamp-pumped dye laser oscillator: linewidth analysis and injection of amplifier system *Appl. Opt.* **26** 2567–71
Duarte F J, Ehrlich J J, Patterson S P, Russell S D and Adams J E 1988 Linewidth instabilities in narrow-linewidth flashlamp-pumped dye laser oscillators *Appl. Opt.* **27** 843–6
Duarte F J, Liao L S and Vaeth K M 2005 Coherence characteristics of electrically excited tandem organic light-emitting diodes *Opt. Lett.* **30** 3072–4
Duarte F J, Liao L S, Vaeth K M and Miller A M 2006 Widely tunable green laser emission using the coumarin 545 tetramethyl dye as the gain medium *J. Opt. A: Pure Appl. Opt.* **8** 172–4
Duarte F J and Piper J A 1984 Narrow linewidth high prf copper laser-pumped dye-laser oscillators *Appl. Opt.* **23** 1391–4
Duarte F J and Vaeth K M 2018 Electrically-pumped organic semiconductor laser emission *Organic Lasers and Organic Photonics* ed F J Duarte (Bristol: IOP Publishing) ch 11
Everett P N 1989 300-Watt dye laser for experimental site *Proc. of the Int. Conf. on Lasers'88* R C Sze and F J Duarte (McLean, VA: STS Press) pp 404–9
Feynman R P and Hibbs A R 1965 *Quantum Mechanics and Path Integrals* (New York: McGraw-Hill)
Feynman R P, Leighton R B and Sands M 1965 *The Feynman Lectures on Physics* vol III (Reading, MA: Addison-Wesley)
Haken H 1969 *Light and Matter* (Berlin: Springer)
Lamb W E 1964 Theory of an optical maser *Phys. Rev.* **134** A1429–50
Lamb W E 1987 Schrödingers's cat *Paul Adrien Maurice Dirac* ed B N Kursunoglu and E P Wigner (Cambridge: Cambridge University) ch 21
Lemaître A G 1931 A homogeneous universe of constant mass and increasing radius accounting for the radial velocity of extra-galactic nebulae *Month. Not. R. Astron. Soc.* **91** 483–90
Lucianetti A, Janulewicz K A, Kroemer R, Priebe G, Tümmler J, Sandner W, Nickless P V and Redkorechev V I 2004 Transverse spatial coherence of a transient nickellike silver soft-x-ray laser pumped by a single picosecond laser pulse *Opt. Lett.* **29** 881–83
Michelson A A 1927 *Studies in Optics* (Chicago, IL: University of Chicago)
Murray J R 1987 Lasers for spectroscopy *Laser Spectroscopy and its Applications* ed L J Radziemski, R W Solarz and J A Paisner (New York: Marcel Dekker)
Pacala T J, McDermid I S and Laudenslager J B 1984 Single-longitudinal-mode operation of a XeCl *Appl. Phys. Lett.* **45** 507–9
Primmerman C A, Murphy D V, Page D A, Zollars B G and Barclay H T 1991 Compensation of atmospheric optical distortion using a synthetic beacon *Nature* **353** 141–3
Pryce M L H and Ward J C 1947 Angular correlation effects with annihilation radiation *Nature* **160** 435
Robertson J K 1955 *Introduction to Optics: Geometrical and Physical* (New York: Van Nostrand)

Sargent M, Scully M O and Lamb W E 1974 *Laser Phys.* (Boulder, CO: Westview Press)

Saxena K, Mehta D S, Srivastava R and Kamalasanan M N 2006 Spatial coherence properties of electroluminescence from Alq_3-based organic light emitting diodes *Appl. Phys. Lett.* **89** 061124

Shimkaveg G M *et al* 1992 X-ray laser coherence experiments in neon-like yttrium *Proc. of The Int. Conf. on Lasers'91* F J Duarte and D G Harris (Mc Lean, VA: STS Press) pp 84–92

Silfvast W T 2003 *Laser Fundamentals* (Cambridge: Cambridge University)

Steel W H 1967 *Interferometry* (Cambridge: Cambridge University)

Thompson B J and Wolf E 1957 Two beam interference with partially coherent light *J. Opt. Soc. Am.* **47** 895–902

Trebes J E *et al* 1992 Measurements of spatial coherence of a soft x-ray laser *Phys. Rev. Lett.* **68** 588–91

van Kampen N G 1988 Ten theorems about quantum mechanical measurements *Physica* A **153** 97–113

Ward J C 1949 *Some Properties of the Elementary Particles* (Oxford: Oxford University)

Index

www.ingramcontent.com/pod-product-compliance
Ingram Content Group UK Ltd.
Pitfield, Milton Keynes, MK11 3LW, UK
UKHW051854100926
8556IPUK00005B/626

9 780750 3340